At the Edge of the Universe

AT THE EDGE OF THE UNIVERSE

edited by

Simon Campbell-Jones

UNIVERSE BOOKS
New York

This book is not sponsored by, connected to or associated with the NOVA series produced by WGBH.

Published in the United States of America in 1983
by Universe Books
381 Park Avenue South, New York, N.Y. 10016

ISBN 0-87663-433-1

L C 83-6680

83 84 85 86 87 10 9 8 7 6 5 4 3 2 1

Printed in Great Britain
by Mackays of Chatham Ltd

Contents

Introduction

Simon Campbell-Jones

The desire to know is phenomenal. Humans have always been intensely curious, in particular about the sky. Whatever is going on up there, out of our reach, over our heads, out among the stars, has always been especially attractive. And yet, despite its complexity, it has never been assumed to be beyond comprehension. Every age and every culture has had ideas, myths or positive beliefs on the subject. This consuming interest has much to do with two things: human destiny and human origins.

The Greek sky was peopled by a variety of gods with human characteristics, who delivered help or havoc down to earth. The early Judaeo-Christian God controlled affairs from 'up' there in heaven (although nowadays he is seen as being less precisely located). Temples, churches and altars around the world testify to the human need to have someone else in overall control. It is clearly very difficult to accept that we might be in sole charge of our destinies, that there might be no superior beings in the entire Universe. Some people, while rejecting the standard religions of the globe, none the less turn to other extraterrestrial forces, flying saucers, interplanetary or intergalactic ancestral visitations, or the ancient and widely followed system of astrology. All these influences controlling our future are seen as descending upon us from the sky – out of the blue, as it were. And if they control our future, they must have controlled our past.

So, inevitably, questions of destiny are bound up with questions of origin. Where do humans come from? Where did Earth come from? Where did the planets, the Sun, the stars, the galaxies, the quasars, the pulsars and all the rest come from? Again every culture has its creation myths, from turtles and lumpy milk to Adam and Eve and the six days that made the world. The modern myths stem almost entirely from the work of astronomers and astrophysicists. Unlike previous natural philosophers they are able to back up their answers with incontrovertible evidence. The speed of light can be physically measured. To look out into space is to look back in time. Light from the Sun takes about five minutes to get here so we see it as it was, and where it was, five minutes ago. Mars appears as it was twenty minutes ago, Pluto eleven hours. The nearest star, Alpha Centauri, is four years and four months away (4.3 light years), the nearest galaxy 2.5 million years. So when the scientists say they can measure the distance to the edge of the Universe, they are also talking about how far back in time they can see – almost, they think, to the origin of the Universe itself.

It is the general fascination with this subject on the part of all of us, which enables so much time, money and effort to be put into it. The building of giant telescopes and the launching of rockets, satellites, shuttles, spacecraft – both manned and robotic – to explore local and distant parts of the Universe is not cheap. But neither is it the most expensive of scientific activities: weapons and nuclear physics are well ahead.

Not least among those who share this fascination with celestial phenomena and the origins of things in general, are the science journalists and television producers. The plethora of information pouring from astrophysical laboratories over the last decade or two has led to a plethora of television programmes. Although it is a notoriously difficult subject to film (most of the evidence is in the form of fuzzy photographs or computer print-outs), producers will return again and again to the topic. The following chapters are based on programmes made in the BBC-TV series, Horizon, *in the hope that some of the producers' enthusiasm for the subject will find an echo with the readers in the same way that it seemed to do with the several millions of viewers who watched the original programmes.*

Our plan is to take the reader on a journey from the surface of this planet out to the edge of the Universe. During the brief voyage there will be many visits to individual planets and star systems. In this way we hope to build up a comprehensible, if not comprehensive, portrait of the Universe – so far as it is known.

The first step in the journey to the edge of the Universe is the one which lifts us off this planet. The story of how this was done by the Russians and the Americans during the 1960s is well known. The first 'giant leap' on to the Moon was not only amply recorded but was watched by millions of television viewers as it happened. What is less known is the outcome of various experiments in space which were carried out by the Americans in the 1970s, after the manned explorations of the Moon. Although the Russians frequently extend their record for the longest period that a man has spent in orbit, the information which they glean, if any, is not available. It seems possible that much of this repetitive circling is for internal propaganda purposes. It was therefore one American experiment, designed largely to study the performance of humans in long space flights, which may have changed the course of future space exploration.

This was 'Project Skylab' which as long ago as 1973 placed in orbit the biggest spacecraft ever built. Known irreverently as 'the ol' tin can in the sky' because its central section was made from an unwanted rocket fuel tank, it was inhabited for a total of 171 days. Three separate groups of men lived, worked, ate and slept inside its fragile skin under medical and psychological monitoring of unprecedented detail. Although Skylab only really captured public imagination when it disintegrated over Australia in 1979, the project had long since engaged scientific interest as NASA's most sustained experimental effort to understand the physiology of man in space. The results of that experiment are now available and they tell a fascinating story: but the basic enigma remains – just what does happen to the human body in a situation where what goes up does not necessarily come down?

1 Zero g

Stuart Harris

Gravity: we are what we are because of it

Gravity does more than just keep our feet on the floor and make things fall downwards. The attraction the planet has for everything on its surface has profoundly influenced not only our technology but the whole course of evolution. To an alien intelligence, the structure of man's body would be a good give-away of the size of the planet on which he lives. Man has a skeleton strong enough to hold him upright; muscles strong enough to make that skeleton run fast enough to escape danger; a heart strong enough to pump blood up his body against the pull of gravity, and gravity sensors in his ears that tell him whether or not he is upright. A champion polevaulter might regard gravity as something which limits his achievements, but in a structural sense it makes him what he is.

Man has evolved to suit his planet in several ways. He breathes its atmosphere, eats its organisms and drinks its most plentiful liquid. So, when his sense of adventure makes him build rockets and leave the planet, he will be going into an environment which lacks everything for which evolution has equipped him, into a place devoid of air, water and gravity. Though astronauts can and do take enough of their home planet's air and water with them to survive, up to now it has not been possible to take along a supply of Earth gravity.

Gravity, Astronauts and Orbits

It is not that gravity is absent in space – on the contrary, Earth gravity is what *keeps* a spacecraft in orbit and stops it going off at a tangent. One way to understand an orbit, is first to imagine an astronaut standing on a fixed platform a hundred miles or so above Earth. The support the platform provides in resisting the pull of gravity gives him a familiar sense of his own weight. He

Skylab 1 in Earth orbit, photographed from Skylab 2 Command/Service module, 22 June 1973.

feels gravity. Now suppose he were to be dislodged. During the fall he would feel weightless because there is nothing to resist gravity – though the outcome of this particular spaceflight would probably spoil the fun. However, if the astronaut is somehow pushed sideways with enough force, he escapes Earth gravity altogether – and that is the last we see of him. But if he has not enough sideways speed to escape, he falls towards Earth – but misses it. The force of gravity, always trying to pull him in the direction of Earth, remains roughly at right angles to whatever direction he's going. So he will fall towards Earth forever without getting any closer; he will be in orbit. *That* is why there's no *sensation* of gravity in an orbiting spacecraft and an astronaut inside it is free to enjoy the micro-world of weightlessness, performing gymnastics that would be impossible anywhere on Earth.

In their passion for being ruthlessly educational the Skylab astronauts used their own bodies to demonstrate certain laws of mechanics. One demonstration they all enjoyed, and which was most entertaining to watch on the Skylab TV, was to suspend themselves in mid-air and then (unsuccessfully) try bodily contortions to 'swim' back to the spacecraft walls. No matter how an astronaut twisted and turned, his centre of gravity stayed in precisely the same spot – and this demonstrates the principle that a closed system cannot achieve locomotion without something to push against. Those with a taste for space drama will note that this also demonstrates that an astronaut placed carefully far from the walls of a space station would be marooned in an ocean of unresisting space – prison without bars.

Fun and Games in Zero g

Because they have to live through a variety of gravitational experiments, astronauts refer to normal Earth gravity as 'one g'. While the peak acceleration of a rocket submits them to nearly three g, total lack of gravity is known as zero g. Astronauts have always exploited the humorous potential of this state and the Skylab crews carried on the tradition of horseplay – but with a new dimension added by the sheer volume they had in which to play games.

For example, the curve of an object in flight on Earth is, of course, a feature of gravity while in zero g a dart thrown at a target or a ball thrown at a fellow astronaut goes dead straight. Things behaving in this otherworldly way are surprisingly hard to catch and the fun, as that most enthusiastic ball-thrower, Commander Paul Weitz, Pilot of the first Skylab, explained, was to make the catch as difficult as possible, but possible all the same. To test their skills they also invented simple games, such as seeing how many bounces they could get when they threw a ball against the ring lockers round the upper part of the workshop. Another was to throw a ball from the midpoint near the airlock down towards the aft end of the stack and try to make it hit the trash airlock in such a way as it would bounce back the length of some seventy feet and up into the command module.

The astronauts also had fun with all kinds of other objects which, deprived of gravity, were dominated by the weaker forces which gravity normally swamps. A gyroscope, set spinning fast, stays in mid-air and keeps running for days. Gerry Carr, Commander of the third crew, seized this unique opportunity to make a short instructional film to demonstrate the curious and useful properties of the gyro. In the absence of gravity and friction, a gyro will continue to point in the same direction no matter what happens to the attitude of the spacecraft. It becomes a sort of astrocompass.

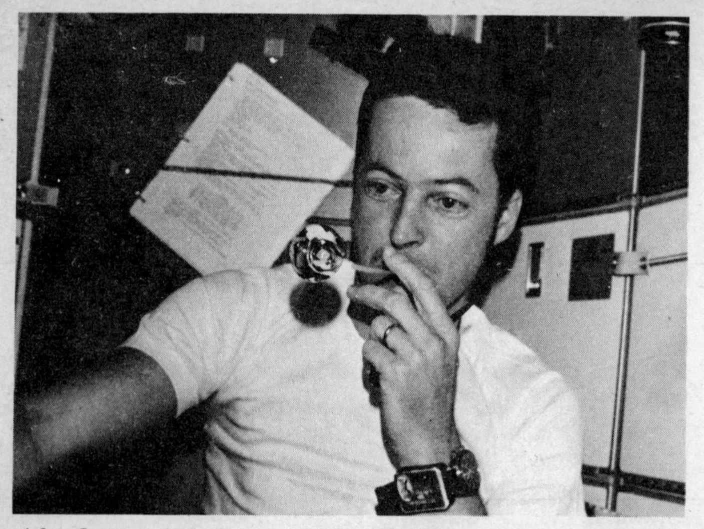

A free-floating sphere of water: experience during Skylab simulation training, March 1972.

The behaviour of liquids was equally instructive and certainly the most intriguing to watch. In zero g any freefloating body of water eventually takes the shape of a perfect sphere because the forces of surface tension are trying to reduce the surface area to its minimum. When attached to something solid, it becomes a hemisphere with something of the properties of jelly. Water was probably the most constantly used Skylab toy. The astronauts inflated water on drinking straws, stretched it into thin sheets, even improvised a lathe to twist fluid into a rope.

Bill Pogue, third crew Pilot, succeeded in proving that contrary to NASA official doctrine it *is* possible to drink normally in space instead of sucking a tube. Using the principle of surface tension, he attached a sphere of water to an improvised cup and, with the TV camera trained on him, took the first ever normal-looking drink in space.

Work in Space

The Skylab crews were not, however, put in space just to fool around in zero g. They had to earn their keep with a range of science experiments requiring sustained mental and physical effort. Lack of gravity can make light of work, too – especially when it comes to handling massive but weightless equipment.

Some of the observation equipment *was* massive, and had to be rigged and de-rigged often enough for the crews to develop their own techniques of throwing things around in space. Al Bean, Commander of the second crew, gave a TV demonstration of his special method of spinning equipment into place, remarking that if the scientists could see what he was doing with their expensive hardware (which of course they could) they would think he was crazy. It looked as if he was being careless but he was not. Heavy items stay in the position they are put – even in mid-air, of course. But if you want to see the other side, you just give it a spin. It moves fast and easily and never seems to get out of control. You can wander off and leave it for a while: it won't float away.

But small objects tend not to stay where you put them. The astronauts complained that Skylab designers had not thought enough about small things like nuts and bolts floating away, which they did – constantly. 'Just put something small down for a second and it's gone,' said Bean, as a warning to the next crew. In one way it was rather useful when working on some detailed repair task just to hang a screwdriver anywhere convenient in mid-air until it was needed. But it had to be placed very carefully: the slightest push as you let go would send it spinning away. All lost objects in Skylab would slowly drift in the air-circulation stream and be found, sooner or later, attached to the screen over the air outlet trunking. This fact, soon discovered by the astronauts, led to the design of an air-current workbench for zero g maintenance.

EVA

Work outside in a spacesuit – EVA in astronauts' jargon (standing for Extra-Vehicular Activity) – has always been difficult, and on Skylab they had to do far more of it than planned, thanks to an accident at launch that threatened to deprive Skylab of electric power. The two main solar panels should have stuck out like aircraft wings and supplied two kilowatts of power each. When the first crew rendezvoused with Skylab, they discovered one panel missing altogether and the other prevented from extending properly by debris from the accident. There was therefore no alternative but to get out and free the jammed main panel with improvised cutters on the end of a pole. The difficulty of EVA work in zero g was evident from the conversation between Pete

Skylab EVA training in a neutral buoyancy tank.

Conrad, suspended precariously in space, and Rusty Schweickart in Mission Control in Houston who had rehearsed the procedure in a water tank.

CONRAD: (*To Paul Weitz, working with him outside the spacecraft*). You pull on the right rope while I'm holding it in place, cause I . . . I can't do both.
WEITZ: Just a second . . . OK, I've got the right rope.
CONRAD: That's the one . . . (*breathing heavily*)
SCHWEICKART: Just for your information, we operated on the opposite side of the discone from the one you're operating on. That is, we operated from the *right*-hand side of the discone. That may help if you need more pole.
CONRAD: It's not a question of *pole*; I've got more than enough pole, Rusty; it's a question of keeping my feet from flying away . . .
SCHWEICKART: OK, the only thing I can say [is] that in the water tank we stood up almost parallel with the discone

with our feet down by the base, and used the discone as a handhold. That helped us – you might want to try that.

CONRAD (*decidedly irritated*): Yeah, I'm doing that. It's not a hand hold I need, Rusty – it's a *foot*hold.

SCHWEICKART: Right. We put our feet right at the base of the discone . . .

CONRAD (*cutting him short*): That's where they are, Rusty.

SCHWEICKART (*Giving up*): OK.

But they finally succeeded – and the third crew accomplished an even more difficult task in EVA when they erected an emergency sunshade to take the place of a shield also torn off at launch. It made the astronauts realise how much one uses gravity to provide a friction hold when one is working down on Earth.

Zero g Housekeeping

The whole Skylab assembly was 117 feet long and weighed nearly ten tons. An airlock was built in so that during EVA the smaller end could be depressurised without wasting precious air from the huge workshop area, which was divided in two by a main floor. The crew boots had cleats on their soles, which fitted the triangular grid on the floors, so they could hold themselves down in any position at any angle. Beneath the main floor was the last section, 6 feet 6 inches high. Known as the wardroom, it was here that the crews felt most 'at home', sleeping, washing, housekeeping and eating in zero g. Not so much a bed-sit as a 'bed-float'.

They had preselected their own six-day cycle of menus for the communal meals, dispensed in the correct sequence from drawers or, if necessary, a fridge. The only major problems with Skylab eating was the need to treat the whole business as part of an experiment for the benefit of the mission biomedical team. No departure from the cycle was allowed; one astronaut is said to have remarked late in his mission, 'I've asked myself every six days, "How come I picked beef hash for breakfast?"' But Skylab crews were the first American astronauts to eat recognisable food with proper cutlery, and they appreciated it, though they said food tends to taste bland up there. It was some time before they came up with the idea of using a hypodermic syringe to squirt salt solution into the food. Salt cellars have to be banned

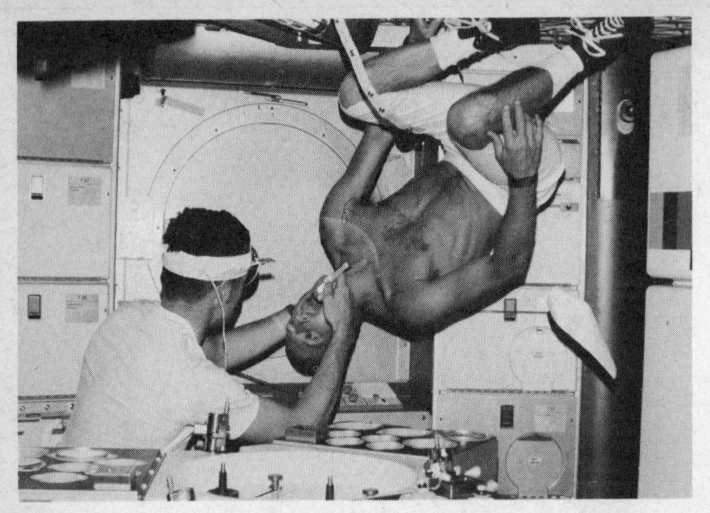

Medical examination aboard Skylab 1/2. The 'patient', Astronaut Charles Conrad Jr, stands on his head, held only by a leg-restraint.

in zero g – salt gets everywhere. The nutrition rules, too, required them to recan any left-overs, make a note of the weight and report to the ground. This was a somewhat tricky process, as you couldn't just scrape your plate off into a bag. This had the effect desired by the nutritionists – it was generally felt to be simpler to just eat it all up like a good boy.

Spacecraft designers find it hard to imagine zero g life so they listen hard to astronaut criticisms. Some minor faults in the Skylab wardroom were noted by Bob Bond, an expert on what NASA calls 'spacecraft habitability'. One of the design aspects he thought might be easily changed in future spacecraft was the horizontal surface of the food trays. It was designed flat like an ordinary dining table. This forced the crewmen to crane their necks and lean over the table to reach objects further away. It would be better if it was angled towards the crewmen, so they could see and reach all the elements on it. On Earth all these things would fall on the floor. In zero g there is no need to have the food far away from the face: in fact it is a disadvantage, and many crewmen picked up the cans, placed them in front of the face and ate, moving the food only a short distance.

Another problem was that some food had to be rehydrated with hot or cold water available at the table. To do this the con-

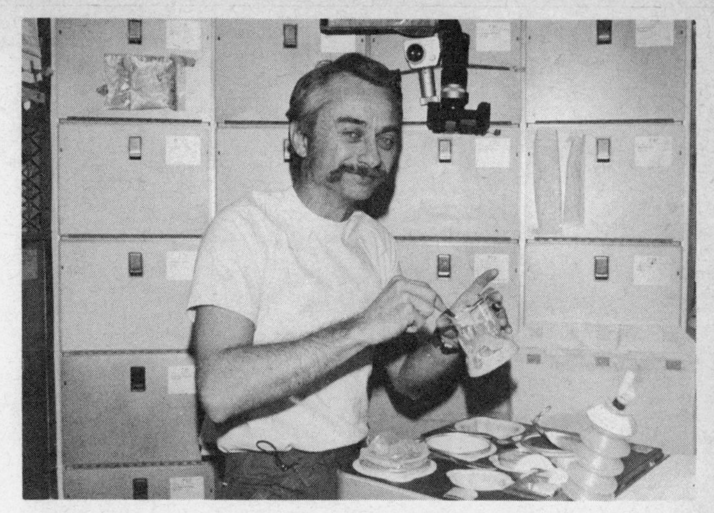

Skylab 3 Scientist-astronaut Owen K. Garriott mixing a pre-packaged container of food in the crew quarters wardroom of the Orbital Workshop.

tainers had to be forcibly pushed down over a spigot and then a valve activated by twisting. The astronauts found themselves spiralling away from the table and, in order to restrain themselves, had to attach their feet to the floor. So you cannot design anything for use in space that requires gravity to hold you or the object in position. Even using a screwdriver you have to hold on to something or you spin round and float away.

In every detail of their daily lives the crews had to restrain themselves, in a physical sense. Paul Weitz, the great enthusiast for zero g sports, found some of the more trivial aspects of life in zero g frustrating. You do not realise just how useful gravity is until you consider so simple an operation as washing your hands under a tap. As Weitz remarked, 'Gravity conveniently carries the water over your hands and out the drain.' He found it a little annoying to be deprived of this through weightlessness.

Skylab's personal hygiene compartment was, though, quite ingenious. There was a little mechanical face cloth squeezer which vacuumed the water away, special electric razors so designed as not to fill the whole spacecraft with floating fragments of whisker, a magnetic soap-holder and suction hooks and velcro pads all over the place to keep control of towels and toothbrushes.

So far as the more intimate problem of defecation in space is concerned, to compensate for the absence of gravity which drops the faeces into the bowl of the water-closet, a blower device sucked it into a bag. In fact body wastes, just like food wastes, had to be wrapped, carefully weighed and returned later to the ground medical team for analysis.

Waste that did not have to be brought home – like dirty laundry – was disposed of through an airlock into a large tank designed for the purpose. NASA definitely did *not* want to be accused of polluting space with the half ton of old socks and used tissues that ended up in that tank.

The astronauts slept in padded zip-up bags attached to the walls of individual compartments – with straps for those who wanted a familiar sense of pressure against a bed. 'As long as there was pressure somewhere on your body, you could convince yourself that you were laying in a bed,' said Ed Gibson. 'I never had any problem even the first night sleeping in the command module. They usually worked us hard enough that we had no problem.' It may sound uncomfortable to be hung on the wall, but there is no 'up' in zero g; and Al Bean actually slept 'upside down' to avoid, he said, a draught up his nose from the ventilation. In fact, without gravity there was in theory no real need for a bed. Just relax and you'll float. But Gibson discovered that the snag to that was that 'When you relax, and float free, it's like in a swimming pool, where your arms are out. You're just in a very relaxed position, and occasionally you would hit a wall; very slowly, but you'd still hit one, and then your mind would say – "I know I'm going to hit another one some time; when's it going to happen?" You'd never get to sleep that way!'

Spacecraft Architecture

What Ed Gibson called a wall might just as well be called a floor or a ceiling – such words are the language of one g and have no obvious meaning in space. Freed from the constraint of a natural 'up' and 'down', designers have to decide whether to give a spacecraft an artificial 'up', and if so – which way?

Decisions like this are part of what Bob Bond calls, rather grandly, 'Spacecraft Architecture . . . the manner in which the vehicle has been laid out and constructed, and the interfaces that people flying in that vehicle will have with it.' The problem was that the astronauts had to train for their flights here on Earth – so the main compartments of the spacecraft had to be usable in

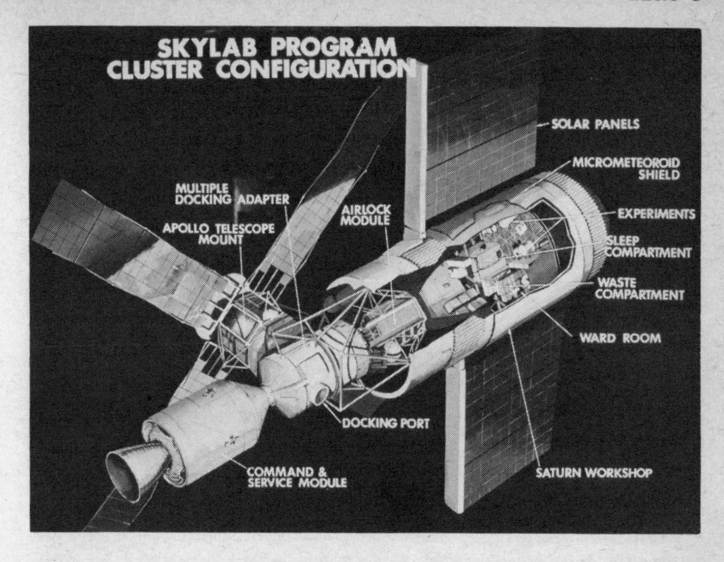

Earth gravity – and the crewmen acquired during training a sense of what was 'up' and what was 'down'.

But Skylab had five different compartments and 'up' did not mean the same everywhere once the whole thing was assembled in Space. At one end of the assembly was the Command Module, which ferried the crews to and from Earth. Here 'up' meant towards the middle of the assembly. At the other end, in the workshop and wardroom, 'up' was also towards the middle – an opposite 'up'. Between them were the airlock and Multiple Docking Adaptor – both cylinders without any obvious 'up'. The MDA in particular was a true wrap-around world, with equipment and even labels oriented every which way. The crews were not trained to regard any particular direction as 'up' in the MDA, and Paul Weitz admitted that going into it from the workshop caused him momentary confusion.

Ed Gibson, though, did not find that a problem. He discovered that to know the vertical of the whole room was unnecessary: you only had to worry about local orientation. If, on Earth, you had papers on a desk arranged at different angles, it would be hard to know where to sit. If they are all orientated towards you, there is no problem – even if a desk on the other side of the room is orientated in a different direction. Weitz, however, insisted that, for him, wrap-around architecture might be logical, but it was not easy to relate to. In fact, he felt it 'better to accommodate people in what they are accustomed to. Therefore I think that there ought to be, within a given compartment anyway, an "up", as it were.'

The spacecraft 'architects' had expected the astronauts to move around mostly head first, as a goldfish does in a bowl, but analysis of the way they used spaces showed that where there was a training-acquired sense of 'up' they tended to stick to it. The explanation was simple. Face to face relationships are unsatisfactory if one of the faces is sideways or upside down. Besides, in the workshop at any rate, the labelling on the equipment and lockers was oriented to be read conventionally – another inevitable carry-over from their long training. There were nineteen thousand five hundred items stored somewhere in Skylab and they all had a place. You simply cannot leave things lying around in zero g. The trouble was that in spite of their naval discipline, the crews constantly misplaced things; they blamed the designers for an illogical locker-numbering system.

Another design problem was communications. Voices only carried a few feet in Skylab's oxygen-rich atmosphere at one-third Earth pressure. So there were intercom boxes in twelve locations but the crews found that if you turned the volume up loud enough to hear properly the loudspeaker squawked. Al Bean gave the next crew – and the design team – the benefit of his opinion that the intercom was 'a great theory on the ground and it looks great in the lab, but up here it isn't worth a darn.' His evaluation was duly noted and stored in a vast filing system under *'Habitability – subsection: communications problems'*.

Spacecraft of the Future

The idea was, of course, to feed the contents of that filing system into the brains of those designing future manned spacecraft, but NASA's dwindling financial resources soon saw to it that there was only going to be one more manned spacecraft design in this generation – the Space Shuttle which finally flew in April 1981 after a delay of three years and a cost overrun of 20 per cent.

With cost-effectiveness dictating that payload gets the lion's share of the available space, Shuttle's living quarters are a step backwards in terms of habitability. Shuttle may therefore be a little too economic for the comfort of up to seven people staying in orbit for up to 30 days. It is true that the Skylab recommendation for an adjustable working and eating surface was picked up, but there is little else to show for all that patient cataloguing of astronaut criticism. The living space underneath the flight desk is just too small ever to become an astronaut's zero g dream world. And one of Skylab's great successes – the zero g toilet facilities – had to be changed anyway for Shuttle because they will be flying mixed-sex crews. What Chris Perner, one of the designers, terms 'the waste management compartment' on the Shuttle is therefore simply screened off by a privacy curtain. Urination is no real problem since this is effected by an adjustable cuff which can be removed for cleansing or replacement. So, although they have cracked the problem of how to spend a penny in space, America is no longer spending enough on manned space flight these days to continue the much more fundamental research into the problems of human physiology – what actually happens inside the human body in zero g.

Space Queasiness

Doing without the downward force which has so dominated our evolution and way of life can at first be unpleasant . The effect of zero g, noticed on Skylab as much as on previous manned missions, is what is known technically as 'vestibular disturbance'. To the victim it feels just like motion sickness, but it is no help to tell him that no motion is involved. Five of the nine Skylab crew suffered some symptoms, the worst case being that of Bill Pogue, third crew Pilot, who vomited just before docking and was out of action for more than a day. More typical were the reactions of Dr Ed Gibson, Scientist in the same crew. He found that after the first two or three days the feeling left him and, although he was vaguely aware of it for the next two or three weeks, he suffered no discomfort after that.

NASA's Director of Life Sciences, Dr David Winter, quite frankly admits that they do not understand the vestibular problem, since none of the predictive tests conducted on Earth appear to have any value in space. While accurate predictions can be made about individuals' proneness to car-, sea- or air-sickness, the tests appear to have no validity in zero g. In fact people who seemed to be rather sensitive had no problem in space, while some of those most resistant to motion sickness in ground tests have succumbed in space. Motion sickness is a little-understood malfunction of a small part of the inner ear called the otolith, which gives us our sense of balance. But the greater mystery about *space* sickness is that astronauts develop strong resistance to it.

The pre-flight training for vestibular function involved sitting in a chair that was rotating rather fast in a room with black and white zigzag patterns on the walls, keeping your eyes open and going through a sequence of head and neck movements. After about sixty movements, most of them were, as Owen Garriott (another who had to reach for the plastic bag in space) put it, 'ready to say "I've had enough for today" and go on to something else.'

But the astronauts continued to submit to this devastating experiment once they were up in space and their performance improved dramatically. On his first run, six months before his flight, Ed Gibson registered a 'malaise level' of six and gave up the test halfway through. By the time of his last training run, he could stay the course but still felt queasy. But once in space he was consistently able to go the full course without reporting any malaise at all and, still more remarkably, he kept that resistance to motion sickness for at least two months after returning to Earth.

Loss of Blood

The second symptom of zero g was also well known from previous missions. Absence of gravity causes a general shift of body fluid towards the head. 'You get a fullness feeling in your head and your neck like when you hang by your knees from a tree-limb,' said Paul Weitz. Since gravity creates a pressure gradient in the human body which tends to make blood collect at the lower ends, the way your brain feels when you hang from a tree-limb is the way your feet feel all the time – but they are used to it.

In zero g, the blood distributes itself more evenly throughout the body. The feet lose blood, but the chest gets correspondingly more; and receptors in the heart interpret that as an increase in the *overall* volume of blood in the body.

One theory about what happens next is that the heart sets about readjusting by sending a message to a gland in the brain: 'Too much blood!' The brain gland starts releasing less of a hormone which affects the kidneys. The kidneys in turn control urine production, so urination increases and thirst decreases, until the total fluids in the body are reduced to bring the heart back to a situation it considers normal.

In an attempt to counteract this effect Skylab had a sort of vacuum tank which an astronaut could get into up to the waist. It simulated gravity by sucking blood towards his legs. Daily

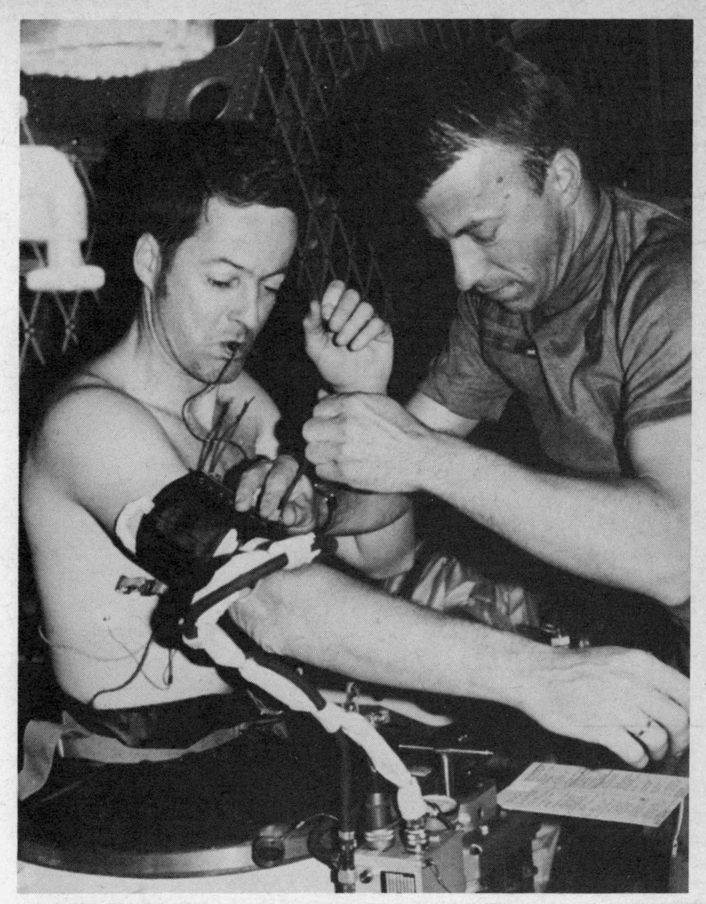

Astronaut Joseph P. Kerwin in Skylab's 'lower body negative pressure device'.

samples of blood and urine were expected to provide the data the physiologists needed to understand the problem of fluid shift completely but, as Dr Winter admitted, in practice, they provided descriptive data without a clue to the mechanism.

For example, consider the levels of the hormone that controls the kidneys, in the case of the second Skylab crew. Just as predicted, the amount decreases progressively over the course of the mission. On the other hand, the picture provided by urine volumes is perplexing. Were they urinating less, or more, or just the same? Little wonder that Dr Winter is reluctant to claim any clear understanding of the body fluid problem.

So long as they get enough exercise, astronauts feel reasonably fit and the reduction in fluid volume appears to have no ill effects – in fact they say it may well help to dispel that 'hanging upside down' feeling. Dr Winter sees it as a logical response to zero g: 'It is basically putting the human body at ease with the new environment. At this point in time we don't see any cause for concern in the cardio-vascular system, or that . . . this new body state is necessarily a bad state. Now that's of course providing you stay in that environment; if you leave the environment then we might have a different circumstance.' Indeed we might, as any astronaut will testify. On coming back to Earth gravity, a smaller volume of blood now falls back into leg veins made slacker and flabbier through lack of exercise.

The heart senses the lack of blood and starts readjusting, but meanwhile there is insufficient blood to go round and the brain, in particular, at the top end of the pressure gradient, tends to feel deprived. This accounts for the fact that after splashdown astronauts tend to need help to get them along the red carpet to a well-earned rest. So fluid loss in zero g *does* matter if you ever want to come back – and especially if it keeps getting worse as missions get longer, suggesting that there is an ultimate limit to the time man can spend in space.

Direct measurement of body volume in space proved difficult, but they could and did measure their body *mass*. The method used to weigh someone is not obvious when there is no gravity to

Alan L. Bean, Skylab 3 Commander, in the oscillating chair.

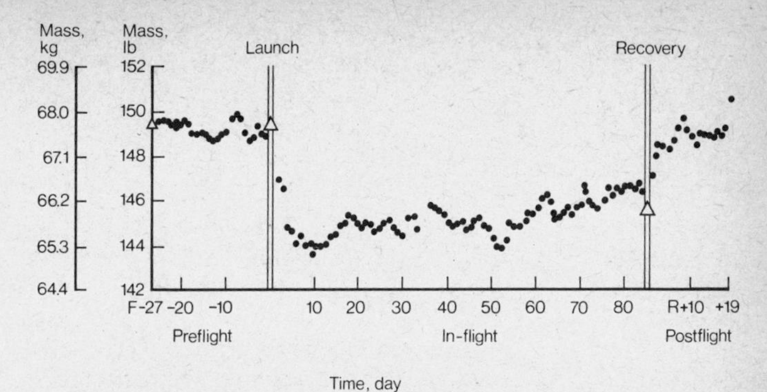

Graph of Bill Pogue's body mass measurement.

pull him onto a weighing machine. They employed the same means used for weighing food – an oscillating chair of which the period of oscillation would get longer the heavier the astronaut was. Up to about fifty days, the trend was downwards all the way. But the chart giving Bill Pogue's mass change during his 84 day mission shows a spontaneous rise during the last 30 days. In all cases, the return to normal after splashdown was very rapid. Mass loss, then, is mostly due to fluid loss, so it looks as though this is self-limiting and that, at worst, an astronaut will lose the equivalent of a wine bottle of fluid from each leg.

Exuberant Blood Cells

Study of the blood samples, carefully centrifuged in space, showed up for the first time a very bizarre effect of weightlessness. Up to 15 per cent of the astronaut's red blood cells began to grow spikes and curious protrusions, almost as if they felt the same sense of exuberance in zero g as the acrobatic astronauts. This change in shape, or morphology, *does* seem to have been progressive yet it leaves such experts as David Winter baffled as to the cause. Although they cannot be absolutely sure, it does not appear to be due to a toxic influence. They do not believe that there has been an increase in red blood cell destruction, nor does a study of the enzymes in the red blood cell membranes show major changes in those enzymes. So, whatever the cause of this change there is no evidence that it is neccessarily damaging to astronauts, although it might be a problem with long-term space travel.

The Disappearing Bones

There may yet be one more factor that will set a limit to the time a man or woman can spend in space. (NASA medics have found no reason to worry about the response of female physiology to zero g, and women will fly regularly in Shuttle.) That one possible limiting factor is the intriguing problem of dissolving bones. This phenomenon, which has been known since the pioneer days of spaceflight, is monitored by measuring calcium in the urine and it is the only serious side-effect which shows no sign of levelling off by the end of the longest spaceflight so far.

As Dr Winter points out, on Earth every single human activity of walking, jumping, climbing up stairs, subjects our bones to stress, even to rather major stress. In zero g this simply does not happen and he believes that it is perhaps this change in the stress of bones which contributes significantly to the calcium loss. Yet however little you may need bones and legs in space, you will certainly need them when you return to Earth. If, then, the astronaut is losing bone tissue at such an appreciable rate, there must be a point in time when an astronaut dare not come home. Given that there are some who consider that if your bones lose about 10 per cent of their calcium they might become very fragile, Dr Winter believes that the time limit varies between eight and twelve months, although the whole issue is somewhat debatable.

With no more long spaceflights in the budget, NASA medics who want to study this problem are reduced to keeping volunteers in bed for long periods. Volunteers flood in at first, but soon become fed up with the boredom and discomfort – as any bedridden patient could have told them. Their loss of calcium in bedrest follows the zero g pattern quite closely. So far there has been no natural reversal of the calcium loss trend, so the real problem is to persuade a healthy person to stay in bed for a year or more and to risk breaking a leg as soon as he gets up.

The physiology of spaceflight, then, seems to be a story of logical adaptation of the human body to a hostile environment, and this is in itself an enigma. Why should a human, without any evolutionary experience of zero g, have *any* ability to adapt to it, let alone such a rapid one? All the experts can say is that other creatures seem to share this inexplicable ability. The Skylab astronauts took with them two spiders, to see how zero gravity affected their ability to spin webs. Though their first efforts were unskilful, they eventually succeeded in spinning reasonable webs.

Minnows in space swim at first in mad circles every which way, demonstrating the need for an 'up' even under water – but they too adapted, by deciding that 'up' was the direction from which light came, and soon swam 'normally'.

Window on the World

Where man has yet to demonstrate his true adaptability to space is in the more profoundly human sense of coping mentally with the very special isolation of long spaceflights. Project Skylab illustrated this problem by the sometimes prickly relationship between the crew and the ground – and by the crew's obsessional Earth-watching. 'If there was one device on board Skylab which was used more than anything else for recreation and interest, it would be our window in the wardroom,' said Owen Garriott, Scientist on the second crew. Skylab's window was rather small and not the best quality optical glass at that. The designers would have preferred not to have one at all – a window is a structural weak point – but the astronauts found it a main attraction, as Ed Gibson explained:

> When we first got into orbit we had our noses pressed against the window . . . we were going across Italy and it was so easily recognisable. . . . For the first couple of days we wanted to keep our eyes against that window. I never tired of that, and found that I could be completely happy doing an Earth Observations mission . . . for ten, twelve hours a day. We found that we got to be quite capable in observing the Earth and we wanted to do it whenever it was possible. For that reason we'd like to see windows that are fairly good-sized, high-quality optical windows from which we can take pictures, . . . throughout the spacecraft in convenient locations, so that regardless of the attitude of the spacecraft we'll always have one window at least pointing towards the ground.

Skylab's orbit took it from as far north as Cornwall to as far south as Cape Horn. The astronauts must have seen every point on Earth between these latitudes.

Their observations and photographs had a practical use as Earth Resources data, but they were *so* insistent on this point, and so unanimous, that NASA psychologists began to wonder if gazing towards their home planet was more than just idle curiosity. Ed Gibson dismissed the idea, saying that he thought that if they were in orbit around Mars he would find himself equally entranced in looking down at Mars, even though he would certainly not consider that home. So, even more forcefully, did Paul Weitz: 'How would you like to go to a cabin in the Alps for two weeks that had no windows, and stay in it all day long: I think it's as simple as that.'

Indeed it is hard to know what part, if any, human psychology will play in limiting the duration of spaceflights. It affects people in different ways, for example, Paul Weitz personally found the effect minimal and felt that he could have gone quite comfortably for as long as a year in space.

But as a Navy man, Weitz did not consider himself an expert on such intangible problems. A scientist, Owen Garriot, was much more prepared to consider the problem; but he, too, rejected psychology as a limiting factor in spaceflight duration. After all people have been going on long sea voyages or wintering in Antarctica for many years now, so there is plenty of experience of small groups in isolation from the rest of humanity. It is assumed, however, that there is interesting or challenging work ahead to sustain attention over the course of the project and with this in mind Garriot himself would welcome the opportunity to take part in such work.

David Winter, the Earthbound expert on space physiology, agrees that a year in space would not cause undue concern. Once the calcium problem is understood, and barring future problems, he foresees an indefinite stay in space as a possibility at least. It might be better if man *did* stay up there indefinitely. The traumas all seem to happen when he decides to come home.

Designs for permanently manned space stations – and there are plenty of them, just no funds to *build* them – usually assume that, apart from playthings like spherical swimming pools in the centre, the living areas will have artificial gravity, created by slowly spinning the whole giant structure (it would still be zero g in the middle of course). But modest-sized space stations with scientific objectives like astronomy will always be zero g microworlds: since a spinning platform makes pointing a telescope needlessly difficult. Furthermore, the trend in NASA budgeting makes it seem likely that if the USA ever does send people to explore other planets, the nonessential cost of creating artificial gravity will not be in the budget.

One way or another, space agencies in the next century may well be recruiting astronauts for very long-term zero g missions.

They may have to be warned that their bodies will adapt – perhaps irreversibly. 'The body does a great job of adjusting to zero gravity,' Ed Gibson reminds us. 'It does just what it should to adjust to the new environment. What we worry about is what happens when you come back.'

To send a manned spacecraft to Mars is quite feasible, technically. The astronauts could be there and back within eighteen months, and there have been longer journeys of exploration. Landing would be no problem, but they would have to take a fairly hefty rocket with them to get off again, as Martian gravity is more than twice that of the Moon. However, the Skylab experience has shown that there may be a serious medical drawback to such a long expedition. Mars is in any case about the limit for manual missions. Who would volunteer to be cooped up, even with good friends, for many years, which is what it would take for a whirl around one of the planets beyond Mars? Think of the food, let alone the inflight movies.

But why does Mars seem most obviously the next stage in our imaginary journey when Venus and Mercury are closer? Unfortunately Mercury is much like the Moon, small, cratered, airless and dead, while on Venus the atmosphere is so thick and hot it would kill you instantly. The Russians, on one of their most scientifically useful ventures, soft-landed a craft on the Venusian surface. As they knew it would, the robot had only a very short life pumping out information before it was fried and corroded into oblivion. Like Venus, Mars is large enough to have an atmosphere, but it is one that might just support life. The job of finding out about conditions on Mars could be done by a couple of remotely controlled robots almost as well as a couple of astronauts – and very much more cheaply. Thanks to computers and microprocessors, two metal earthlings have already stepped on Mars.

2 The Red Planet

Tony Edwards

> We've become quite used in our modern world to tremendous technological miracles. In fact we've become a little bit blasé. And when one mentions the accuracy with which Viking landed, oftentimes one hardly says anything at all about it. But in fact we travelled four hundred million miles to Mars and landed within ten miles of our selected target. That is analogous to someone climbing up on the top of Big Ben with a peashooter in his hand and announcing to his cousin in Los Angeles that the forthcoming pea is going to hit him on his right cheek rather than the left cheek.

These are the words of Gentry Lee, one of the young scientists behind the Viking mission to Mars in July 1976. It was indeed a formidable achievement, the result of years of technological endeavour, but also the culmination of centuries, not so much of science, but science fiction.

Martians

Mars, the red planet, has long held a special place in the minds of men. The ancient astronomer—astrologers of Babylon personified it as the God of War and in our century science fiction writers continued the tradition. Although H. G. Wells and Edgar Rice Burroughs might people the planet with bug-eyed monsters and warlords, threatening Earth with imminent destruction, the most imaginative tales about Mars have come not from literature but from the curiosity of rational men of science.

When modern astronomy was born with the discovery of the solar system, scientists were quick to reject the old astrological notions that the planets were gods, but they still clung to the belief that the planets had some kind of life. They argued that if the Earth was a planet, it followed that the planets too were Earths. So, when Kepler did the calculations for Mars and found its orbit brought it very close to Earth – only thirty-five million miles – it became an obvious candidate for being a world like ours. And, almost uncannily, subsequent discoveries for the next three hundred years repeatedly confirmed that hypothesis.

In 1659 the Dutch astronomer Christian Huygens produced the first map of Mars. Through the primitive telescopes of the day he could just make out that the Martian surface was covered with light and dark patches. Not much to go on, you would think, but he wrote a book called *Cosmotheoros* in which he confidently announced that Mars was inhabited. The first Martians were born.

A hundred years later telescopes got bigger and Mars came a little closer. It not only had those surface markings, but two white spots at each pole. In 1784 Sir William Herschel reported to the Royal Society: 'I have seen the white poles retreat and expand in yearly cycles like the seasonal melting and freezing of ice at the poles of the Earth. I conclude Mars has summers and winters like ours.' In fact Herschel found Mars was even more remarkably earthlike. It had the same tilt in its axis, and it rotated almost exactly once every twenty-four hours. 'I have also noticed changes of bright and dark spots. These we can ascribe to clouds and vapours. Thus the planet has a considerable but moderate atmosphere and the inhabitants probably enjoy a situation in many respects similar to our own.'

Herschel was never doubted. Exploration began in earnest and the first detailed maps were made. Most astronomers seemed to agree that the darker markings on the planet were Martian oceans. It was a reasonable hypothesis, but it had two rather embarrassing problems: such large bodies of water should have reflected sunlight, and they did not. Nor could the astronomers quite agree on the maps – everyone came up with a different one.

In 1877 one astronomer tried a new approach. Instead of looking directly at the surface, Asaph Hall, an American, turned his telescope off to one side of the planet. He was looking for another clue to the type of world Mars was. And he found, circling just three thousand miles above the planet, a tiny moon only twenty-five miles across. A few days later he saw another. Hall called the two moons Phobos and Deimos, names of the mythological acolytes of the God of War.

Now that it had been established as having moons, Mars became even more Earthlike. Furthermore, from the speed of the moons' orbits, the gravity of Mars was calculated to be just under half that of Earth. It was the first real piece of scientific data to come from two centuries of observation.

For the next fifty years Martian exploration was dominated by one man, Percival Lowell, a wealthy American businessman and no mean astronomer. He had predicted the existence of the planet Pluto, but he turned to Mars in the 1890s after seeing drawings by an Italian, Giovanni Schiaparelli, who had found faint linear features on the Martian surface. He called them channels, or in Italian *canali*. By 1900 Lowell had produced his own maps of Mars. He said Schiaparelli's *canali* really were canals, and their geometric and planetwide construction implied the work of rare intelligence. Lowell announced: 'That Mars is inhabited by beings of some sort or other we may consider as certain as it is uncertain what those beings may be.'

Lowell claimed further indirect evidence for the canals from photographs which showed that, as the poles retreated in the summer, a wave of darkening would spread towards the centre of the planet. This, he said, was seasonal vegetation, perhaps crops grown by the Martians, and the water to grow them came down the canals from the poles. As the ice melted every spring, the water was circulated to the dry equatorial regions. Getting the water around, Lowell said, was the major task of a doomed civilisation on a dying planet.

Not everyone agreed: some astronomers never saw the canals. However, there were many who did, and the idea of canals fitted the orthodox view of life on Mars. Only a few dared to suggest that the canals were the product of an overheated imagination at one end of the telescope and the distortion of the Earth's atmosphere at the other. This had been a problem for centuries.

So in 1918, six thousand feet above the haze of Los Angeles, they built on the summit of Mount Wilson the first of a new generation of giant telescopes. At the time, the Mount Wilson instrument was the largest in the world and it quickly attracted astronomers eager to check on Lowell's observations. But it proved disappointing as visually there was little improvement. Nevertheless, the new large telescopes had one major advantage: they increased the amount of light coming from the planet, and with that you could do some proper science. Splitting the light up spectrographically, the scientists were able to analyse its chemical composition and nearly buried the theory of life on Mars for ever. The spectrograph showed the planet had a pretty nasty atmosphere, containing up to thirty times more carbon dioxide than on Earth and only traces of oxygen and water.

Radiation measurements said the temperature ranged between 40 degrees centigrade by day and −70° at night. Indirect evidence suggested a low atmospheric pressure one tenth that of Earth. Intelligent life on Mars was ruled out, but the orthodox view still held that the planet had primitive vegetation.

Mariner

It was perhaps inevitable that with the coming of spaceflight, man's imagination turned towards planets other than our own. While the first man was orbiting Earth, a spacecraft was being assembled to go to Mars. Named 'Mariner 4', it carried a television camera to transmit live pictures back to Earth. In November 1964 it was launched on a journey which required incredible precision. Mariner 4, travelling more than a million miles a day, would take eight months to reach the planet, and sweep by only a few thousand miles above the surface. On 14 July it arrived right on target, but the celebrations were short-lived. The handful of fuzzy black and white pictures it sent back overturned three centuries of exploration. There were no canals, no vegetation, no bug-eyed monsters, just masses of craters. Bruce Murray, Professor of Astronomy at CalTech was there when the pictures came in: 'We were all shocked by seeing such large lunar-like craters. It meant that Mars had not recycled its surface the way the Earth does. There must have been no rainfall, no weathering, in any way comparable to that of Earth for billions of years, in order for Mars to resemble the Moon.' So Mars was not at all Earthlike; it seemed dead like the Moon.

The next two spacecraft they sent confirmed the bad news. The atmosphere was measured and found to be painfully thin: a pressure of only seven millibars – a hundred times less than Earth – so low that any liquid water would instantly boil away. Temperature readings suggested that the polar caps were composed not of ordinary ice but frozen carbon dioxide. Mars became even colder and more hostile. The scientists were disappointed and some of them, like Hal Masursky, could see the political implications too: 'We had already committed ourselves to first a detailed reconnaissance by an orbiter, and finally landers, to look for life. And they wanted to shut that programme off. So as we approached Mars with Mariner 9 there was a great deal of anticipation.'

Mariner 9 was designed to go into orbit around Mars and eventually to map the whole planet. It arrived in 1971 and saw –

Phobos, one of Mars's two tiny moons. The diameter is about 22 km, and the crater at the top measures 5 km across.

nothing. The entire planet was in the grip of a dust storm. So Mariner 9 turned its cameras outwards and caught a distant glimpse of Deimos, the outer moon of Mars. Closer was Phobos, circling three thousand miles above the planet. They looked like pockmarked potatoes and turned out to be even smaller than was thought – Phobos just fifteen miles long and Deimos only eight.

Meanwhile on Mars the dust storm was abating and, as the cameras turned back onto the surface, there appeared four dark spots. The atmosphere cleared further, and the spots turned into the summits of four enormous volcanoes. They were giants, up to twenty miles high and three hundred miles wide. By an extraordinary chance that side of the planet had been seen by none of the previous spacecraft. Bruce Murray could not believe his eyes: 'To find a planet that on one side preserved relics of three or perhaps even four billion years ago in that they were

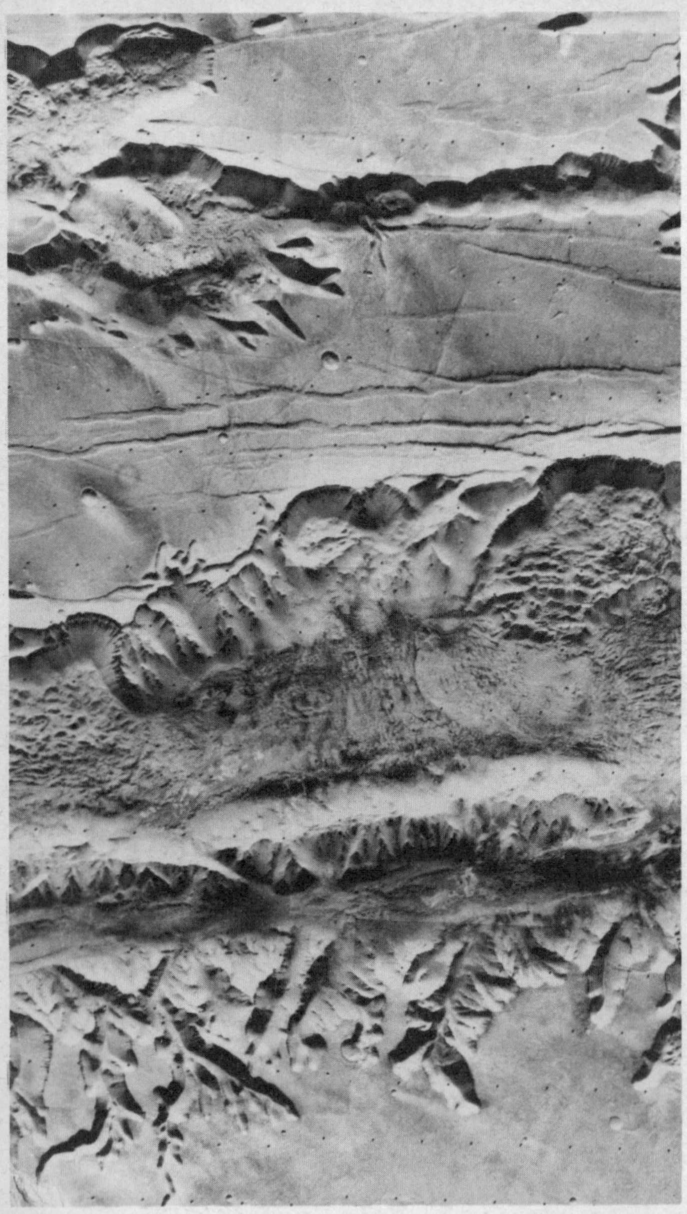

Part of Mars's most spectacular feature, the giant canyon Vallis Marineris, seen from the Viking orbiter.

lunarlike, and yet on the other side was more Earthlike than Earth, was totally unexpected and to many of us the very greatest shock of our scientific career.'

But more shocks were to come. Besides volcanoes, there was a massive scar on the equator that the cameras revealed as a canyon of giant dimensions, three thousand miles long, fifty miles wide and almost three miles deep. There were monster sand dunes too, each ridge a mile apart. They were formed, it was thought, by wind, like dunes on Earth, only on Mars it blew at over one hundred and fifty miles an hour. So here at last was the answer to those changing surface markings that Lowell thought to be vegetation grown by Martians. Mariner 9 saw the changes in close-up and showed they were caused by nothing more exciting than wind-blown dust. So the final hope for life had crumbled.

But not for long: on Mariner 9's three hundred and thirty-first revolution of Mars, it sent back a picture of what looked like a dried-up three-hundred-mile-long river. After that more rivers showed up. They had clearly been caused by liquid. But was it water or lava? For some there was no doubt. Harold Masursky

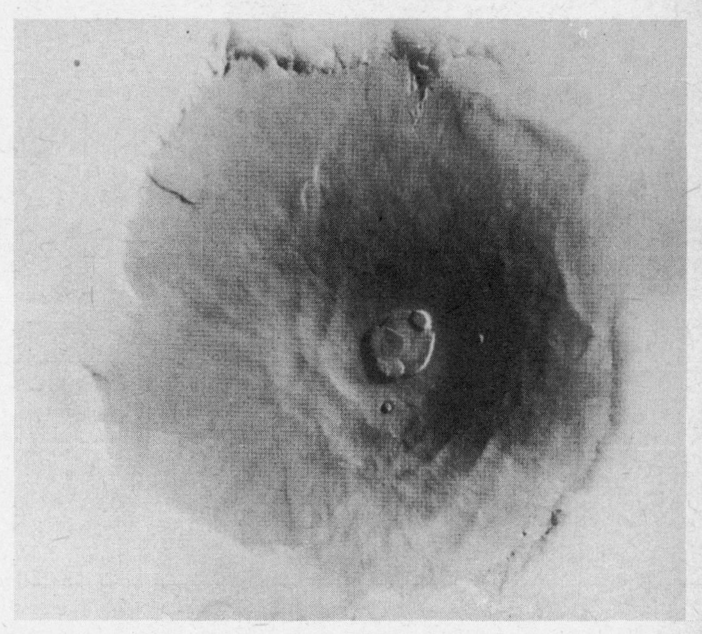

The giant volcano Nix Olympica, 600 km wide and 23 km high: a photomosaic by Mariner 9.

Mariner 9's three-picture mosaic of the 300-mile long channel that appeared to have been formed by water.

echoed the feelings of many of the scientists: 'The presence of the channels meant that the old theories of the dead planet without enough atmosphere to have water were gone. That ugly theory was killed by the beautiful fact of water being there and the chance for life being there.'

The centuries-old dream of life on Mars was reawakened but the scientists still knew it was a long shot. After all, their own instruments had told them the atmosphere was too thin to support liquid water. It was also not thick enough to filter out the harmful ultraviolet rays from the Sun, so life as we know it seemed unlikely. However, the biologists had already begun to design experiments to test the possibility that life in miniature – micro-organisms – may have survived on Mars. One theory was that they were lying dormant and that a cocktail of nutrients would awaken them to life. Another hypothesis was that plantlike cells might have adapted to the harsh ultraviolet light on the planet and survived. These ideas were developed into one of the most sophisticated pieces of hardware ever built, a complete biological laboratory containing three separate experiments designed to work automatically and test the Martian soil for minute traces of microscopic life.

But how to get it up there? A manned expedition was out, being too costly and too dangerous. So instead, they spent fifteen years designing and building a landing craft which would parachute itself down on to the Martian surface and then, alone and unattended, report on the unknown world.

Viking

They called it Viking; it was the most sophisticated robot ever built, a machine modelled on man himself, although in many ways a vast improvement on the human original. First, Viking had two eyes – colour television cameras – capable of seeing stereoscopically almost 360 degrees around the craft. Better than a human eye, they could see in infrared as well. Atop the lander was its ear – far more powerful than our ears of course – the radio antenna link to Earth. Inside the craft another ear, a delicate seismometer capable of perceiving vibrations way beyond a human ear. The sense of touch was provided by the meteorology boom. Sticking up like some giant wetted finger, it measured temperature, humidity and wind speed. The other arm was much longer, it could unroll almost six feet out in front of the lander, pick up soil samples in its metallic hand and return them to the lander's mouth. Inside the mouth, was the stomach, the miniaturised chemical laboratory that processed the soil to discover any living micro-organisms. Another compartment contained the nose – technically a gas chromatograph mass spectrometer – which heated up the soil samples and sniffed at them to see what they were made of.

Of all the humanlike organs which were created for the lander, the most impressive technological achievement was its brain. The brain was of course a computer. It was the size of a suitcase, weighed fifty-two pounds and had a vocabulary of eighteen thousand words. Like a human brain, it really would have to think for itself, especially during the landing. Too far away to be controlled from Earth, the brain was designed to mastermind the descent unaided; it had to sense the landing altitude speed and position, decide when to jettison the heat shield, deploy the parachute and turn on the terminal descent engines.

The Viking lander was probably the most advanced piece of automatic intelligence ever built – a robot made in the image of man – but it ended up looking a bit like a bug-eyed Martian!

The first panorama of Mars seen by the Viking 1 lander (foreground).

Touchdown

In August 1975 the robot attached to an orbiter was despatched to Mars. In fact two landing craft were sent and they arrived within a few weeks of each other the following June, taking up their assigned orbits round the planet. This time the familiar features stood out clearly and Viking 1 took a look at its landing site. From the maps produced by Mariner 9, the scientists had earmarked an area in the plain of Chryse, chosen because it seemed once to have been inundated by a great flood and was thus a likely place for life forms to have begun; it also looked a safe place to land. But the scientists got a shock. 'We were startled by the clarity of the pictures,' confessed Harold Masursky, Landing Site Team Leader. 'We had originally chosen the landing site because it looked very smooth from Mariner 9. But we got these Viking pictures and we could see the site was full of craters, so we decided very soon that there was no way that we could land in our original choice.'

The orbiter was sent off to look for another place to land. Three potential sites were rejected before one was finally chosen, three hundred miles away on the edge of the Chryse Basin. The lander separated from the orbiter and began its descent.

Two hundred million miles away, in Mission Control at the Jet Propulsion Laboratory near Los Angeles, the scientists and engineers held their breath. As the lander descended, it sent back word of its progress; but if the engineers had spotted an error there was nothing they could do. Because of Mars' great

distance radio signals take twenty minutes to arrive – the lander was completely on its own.

Touchdown, we have touchdown.

The control rooms emptied as hundreds of milling scientists embraced each other like so many footballers. The first one to speak for the record was geologist Dr Thomas Mutch (later tragically killed in a climbing accident in 1980). As head of the Lander Imaging Team his task was to interpret the pictures received. Camera 1 had been programmed to look at the lander's left leg. As the television pictures were painstakingly built up line by line Tim Mutch was at first lost for words in the realisation that the incredible had happened – this really was Mars. He hazarded an instant analysis: many of the rocks looked like ones seen in deserts on Earth. They were pockmarked with vesicles, which meant they were probably volcanic.

A few hours later, Camera 2 sent back pictures of the scene in front of the lander – a dramatic panorama of a hostile lonely landscape. 'The most striking impression is one of a lot of rocks,' said Dr Mutch, 'and this automatically brings to mind the fact that we had a good deal of luck, because some of these rocks are about two to three metres across and had the spacecraft landed on those rocks, it would have been probably permanently disabled.' In fact, what Viking had landed on looked liked fine sandy soil; the lander's right leg had got buried in the stuff, and in the distance the geologists could see large dunes. 'It looks so much like a desert environment,' said Dr Mutch, 'you can almost

The first photograph taken on the surface of Mars. On the right is the Viking lander's left foot, planted firmly on the surface strewn with basaltic rocks.

imagine the camels coming up over the dune into our view. Unfortunately we do not see any camels or any other animals for that matter.'

Of course they had not expected to, but nevertheless some of the more optimistic scientists had dreamed of finding primitive vegetation. Meanwhile the rest were laying bets that the soil samples scooped up by the lander's extensible arm would contain microscopic life forms.

By the morning of the eighth day the arm had taken its first grab at the surface, leaving a clear trench in the ground. The biological laboratories digested the soil and returned their analysis to Earth. Across two hundred million miles of space came the message that was hoped to end three centuries of speculation. It came not in the language of men or Martians, but in the chatter of a robot talking to a lifeless computer.

Surprises in the Soil

As the print-outs were scrutinised, biologists like Norman Horowitz confessed to an immediate reaction of astonishment. It became obvious right away that the Martian surface material was very active, and it appeared that organic compounds were being synthesised.

The three separate experiments on board the lander all seemed to confirm the extraordinary news that there was indeed life on Mars. In the labelled release test a solution of nutritious goodies that microbes like to eat had been dropped on to the soil sample. Immediately quantities of carbon dioxide were given off, showing the microbes were happily enjoying their meal. Another cocktail of nutrients was given to a second soil sample and again it appeared to contain living creatures. In a third, the soil sample assimilated carbon dioxide experimentally fed to it, again indicating the presence of life.

It seemed too good to be true, and it was; because a few days later, data which was to cast strong doubt on the results, came in from another Mars-based laboratory, the gas chromatograph mass spectrometer. Its job was to search for organic matter in the soil by analysing what gases were given off when it was heated in a tiny oven the size of a matchbox. At first the soil looked Earth-like; it contained nitrogen, carbon dioxide, argon and traces of oxygen and water, but not a scrap of organic matter. For the biologists it was a bombshell. Their tests had said 'life', and yet life without organic matter was unheard of. Some scientists questioned the spectrometer's sensitivity, while others speculated about cannibal micro-organisms that might live by consuming their own organic debris. However the search for life was not over: there was, after all, the second spacecraft.

Viking Lander Two touched down successfully six weeks later on the other side of the planet and found a rock-strewn landscape almost identical to the first one. Again the soil was analysed for evidence of life, but the results were depressingly similar – apparently very active biologically and yet containing no organic matter. The suspicion arose that if there were no microbes on Mars there must be something pretty odd about the soil. These doubts were strengthened when further tests showed that, at temperatures which would have annihilated all living organisms, the soil still appeared to contain them. After three months Norman Horowitz was ready to admit the worst: 'If the cameras saw a line of trees on the horizon, or if some morning they saw a footprint in the sand which had not been there before, that would be definite proof of life on Mars. The current hypothesis is that the experimental results are in fact signals not of biology but chemistry.'

So the search for life on Mars gave way to another equally intriguing question: what is so strange about its soil? The first clue was its colour. Mars, the fiery spot in the sky, had long been known as the Red Planet. Viking's spectactular colour pictures showed that the surface was indeed a yellowish-brown. To geologists this suggested the presence of compounds of iron, a

The scene in front of the second Viking lander. On the ground can be seen the shroud of the sampler arm, ejected after touchdown.

diagnosis confirmed by x-ray spectrometry from another minilab on board the lander. But what makes this rusty Martian surface so different from iron rich soils on Earth is the action of the Sun. Because the atmosphere on Mars is so thin, the Sun's ultraviolet rays beat down unfiltered on to the soil and are thought to start an unusual chemical process, turning normal iron oxides into superoxides.

This is now believed to be the explanation for the puzzling data from the biological experiments. Add water to superoxides and they start behaving like microbes, giving off gases in profusion. In addition, the super rusty soil itself appears to be extremely hostile to life: back in laboratories on Earth, synthetic Martian soils were found to destroy organic matter.

It has been a disappointment for the biologists, but for geologists like Bruce Murray, Viking was an eye-opener. 'What we're seeing is a new kind of weathering process, a soil chemistry quite different from any existing on Earth. It is this which may help to explain many of the mysteries of the Martian landscape – those huge valleys, pits and channels whose origins simply cannot be explained.'

Landscapes

Mars has indeed turned out to be a geological wonderland. Its massive volcanoes, canyons, lava plains, channels, craters and dunes were seen by the Viking orbiters much more clearly than by Mariner 9, and the sheer wealth of the geological detail astounded the experts. Michael Carr, Head of the Orbiter Team, admitted that much of it was incomprehensible. However they were able to begin to piece together a geological history of the planet.

You can get a rough idea of the age of a particular surface feature by the simple method of counting craters. The more meteorite impacts, the older the area. It turns out that the youngest features are the volcanoes – some of which Michael Carr believes may still be active. In contrast, almost half of the rest of the planet's surface is absolutely peppered with craters; that side seems to have been totally dead for four billion years.

But the real historical puzzle is the origin of the channels – those dried-up river beds first seen by Mariner 9 and now revealed in even greater detail by Viking. Before, a question mark had hung over what had caused them: was it wind, lava or water? Now there was little doubt, the flow lines clearly indicated water.

The orbiters sent back evidence of two distinct types of watery features, vast floods and small rivers. The flood features are in fact immense, showing evidence of what must have been sudden catastrophic events, the water first cutting twenty mile wide channels and then flowing out for hundreds of miles into the plains. Geologists calculate such surface features must have been produced by water travelling at up to thirty thousand million cubic feet per second. This is a vast amount of water for a planet which appears so dry and whose atmosphere is too thin to support liquid water. But the water is there all right. Striking pictures have been sent back of early morning mists and whole banks of clouds covering sometimes a quarter of the planet. That, together with other data, has led geologists to be pretty certain that the water is buried beneath the surface of Mars as a vast layer of ice. The planet appears to be a shell of ice floating in a sea of rock. So the geologists now think they can explain the great floods by volcanic activity that never quite reached the surface. Local geothermal heat melted the ice layer and caused the ground above suddenly to collapse; the water gushed out from underground, instantly eroding a vast channel and eventually flowing out into the plains.

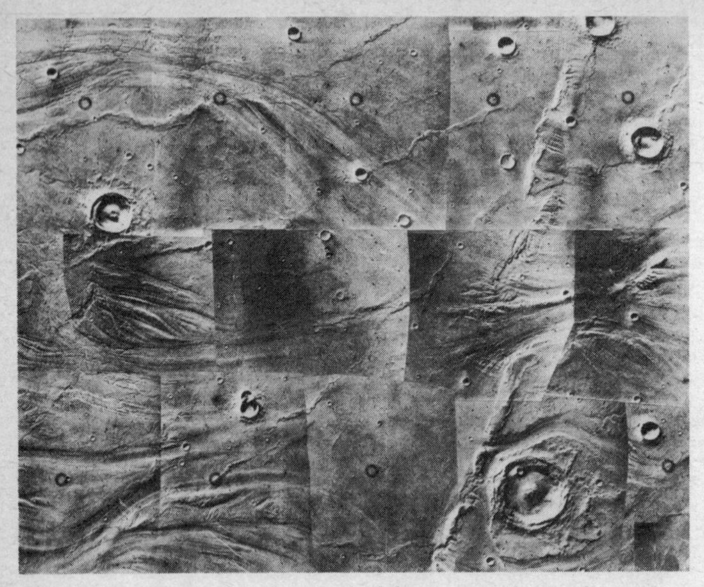

Fifteen-picture mosaic of a 50,000 square kilometre area showing dramatic evidence of flooding.

Far more controversial is the origin of the other water feature on Mars – dried-up rivers. There are two basic types: long channels about half a mile wide, and branching networks of much shorter rivers. Counting the craters in the river beds caused by meteorites is a useful method of comparative dating and this showed that the long channels are up to 2.5 billion years old, but the others are much younger. Were they also caused by sub-surface ice or, as some geologists have thought, by rain? For astonishingly, the evidence seems to suggest that not only were there rainstorms on Mars, but there were also rainstorms at different periods in its history.

The Rain Puzzle

Water on Mars has now become as important as the question of life, but it has raised major theoretical problems. If there were rainstorms at different times, the planet must have had numerous changes of climate and the scientific implications of that are embarrassing. It has led some to question the whole idea. Barney Farmer is one of a handful of British scientists attached to the Viking project and he puts his finger on the problem: 'It's very easy for the geologists to look at pictures and wave their arms and say there must have been rain to account for these small stream-like features. But there is a problem with doing that – you have to have some physics behind it. And the physics of producing rain on Mars is very difficult. Perhaps the physics isn't so difficult as twisting the physics so that you produce rain.'

So far the only hard piece of evidence about Mars' climatic history has come from another Viking experiment and some elegant detective work. By analysing the composition of the isotopes of nitrogen in the Martian atmosphere today, scientists have been able to deduce that in the past Mars did in fact have a dense, Earthlike atmosphere. So in theory it could have once rained on Mars, but only, according to the data from this experiment, in the early part of its history.

And yet the pictorial evidence seems to show the rainlike channels have been formed at a number of different periods. That means the atmosphere has still to be stored somewhere on the planet. But where? The obvious places to look were the coldest places on the planet – the gleaming white poles.

When Mariner 9 the earlier spacecraft had studied the poles it had reported back that they were made not of frozen water as on Earth but frozen gas – carbon dioxide. That was precisely what the scientists wanted to hear: there, they said, was the Martian atmosphere in cold storage. Thus, a slight change in the tilt and orbit of the planet would be enough to heat the poles, release the gas and provide a dense atmosphere where water could stay as a liquid. But this elegant theory was disproved by Viking. Thermal Mapping Team Leader, Hugh Keiffer, announced the bad news: 'Viking's measurements of the north pole temperatures show that the poles are much warmer than is possible for carbon dioxide. It means the poles must in fact be made of ordinary ice – they're not a frozen atmosphere from Mars' past history.'

So the idea that the poles are the remnant of an ancient atmosphere has taken a knock, making any episodes of rain somewhat difficult to explain. In order for the pressure on Mars to reach the level at which water could remain as a liquid, the temperature would have to be raised an improbable 70 degrees centigrade.

Digging Deeper

But the poles have more secrets to reveal. As the Viking orbiters circled a few hundred miles above, they returned stunningly detailed pictures of the icy surface. Down amongst the ice they saw bands of bare ground about ten miles wide and, on close inspection, those bands turned out to contain geological strata. The pictures were seized on excitedly by the geologists who saw that the strata could only have been formed if Mars had indeed had some major changes in the climate. It seemed to confirm the evidence from the watery features elsewhere on the planet and raised everyone's hopes about the original purpose of the mission – finding evidence of life. Even the sceptical Klaus Biemann admitted the possibility that life could have evolved and existed at some periods in the planet's history. 'Those remaining traces of materials would by now be well covered and found only deep underneath the surface in a similar way as we find coal and oil deposits on Earth, which are also remnants of earlier living systems.'

The Viking lander could not dig deeper than a few inches, so that theory could not be tested, but the biologists had one last card to play. Perhaps, they thought, organic matter, living or dead, might still survive on the surface underneath a rock. The destructive ultraviolet light could not reach it there. So, at Mission Headquarters in Pasadena, they put a fully working replica of the lander into a mock-up of the actual scene on Mars. They simulated the soil and duplicated every rock and boulder within reach of the lander. The idea was to test whether the Viking robot had the strength to push one of the rocks aside and scoop up the soil underneath it. It was a tricky business, the lander's arm had never been designed for poking about under boulders; but the test with the mock-up worked, and the command was sent to Mars for Viking to collect the soil sample. The biologists held their breath as the data came down, but they were disappointed – the results were the same as before.

It was the final act in our cultural dream of life on Mars, but Bruce Murray was philosophical about it. 'The search for life on an alien planet was always looked upon, even by the most passionate advocates, as a long shot. The fact that the tests have not yielded confidently positive results implies that the kind of life we were looking for is not there.'

For three hundred years we have been misled by trying to answer the question whether Mars is a world like ours. We now

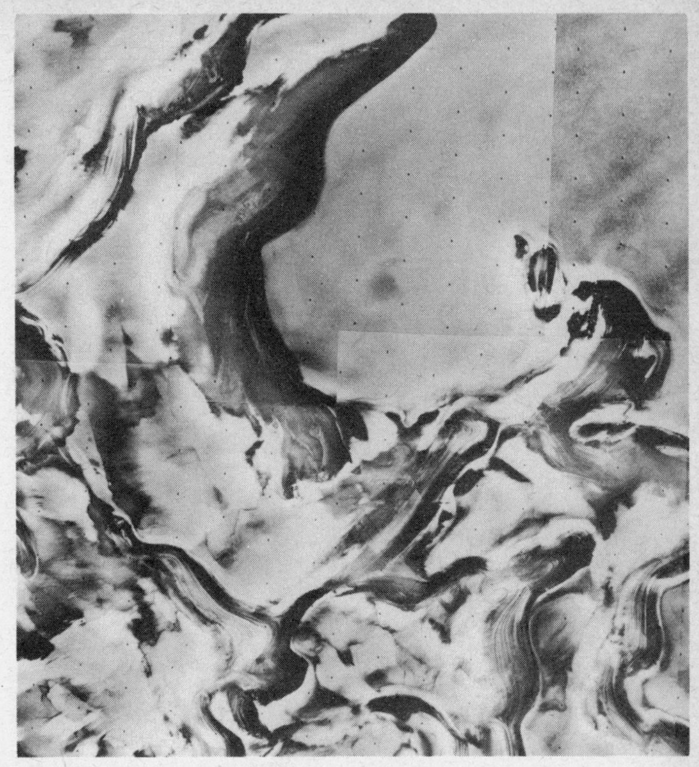

Soil strata at the Martian North Pole. Each layer is about 30 metres thick.

know it is not. It is, in fact, an alien world. It has an alien soil, creating a unique set of geological processes unlike anything on Earth. It is cold, bleak and has a painfully thin atmosphere. And yet Viking has shown that in the distant past Mars *was* probably much more a world like ours. In fact of all the planets in our solar system, Mars was always the one most likely to support life. Now we know it does not, it perhaps will make us appreciate even more the uniqueness of our own planet, our living Mother Earth.

So we must forget about the search for life, if we can, and move on outward. We go through the circular belt of rocks known as the asteroids. Not much is known about their origins: they may be fragments of a 'missing planet' which was shattered in a catastrophic collision or they may be just a collection of cosmic particles that never made it into a planet. The asteroid ring sits very neatly, in geometric terms, in the space between Mars and the next planet, Jupiter. It marks the boundary of a different realm.

The four main outer planets, Jupiter, Saturn, Uranus and Neptune, are huge. The distance between them is huge. Thousands of inner planets could fit inside one of the outer ones. In spite of this size, they rotate very much faster than the inner planets. One Jupiter day is about 10·5 Earth hours, whereas one Venus day is about 243 Earth days. Each outer planet has many moons of varying sizes. At least two, Saturn and Uranus, have their own 'asteroid belts', or rings, but the main difference between these four giants and the four inner planets is that they are made almost entirely of gas. There is no way that any spacecraft is going to land on any of their surfaces. If one tried, it would sink slowly into a sphere of pressurised gas which would behave more and more like a liquid, until the craft found its own buoyancy level, suspended like a bathyscaphe in a deep dark ocean. However, the view from a spacecraft flying by, hundreds of miles above the surface, is quite spectacular enough and, scientifically, more useful.

3 Encounter with Jupiter

Fisher Dilke

Before Voyager

Jupiter is the largest of the planets. To the naked eye it looks like a very bright star. Unlike a bright star, but like the other planets, its position in the sky changes night by night. Its location lies on the ecliptic, the line that crosses the sky intersecting all the constellations of the zodiac. That was probably all that was known about Jupiter until in 1610, Galileo developed an astronomical telescope and became the first man to see more. His telescope was primitive and was about equal in its power to a rather indifferent modern pair of binoculars. Yet it was enough of an improvement over the naked eye for Galileo to see four points of light, like stars, spread out in a line on either side of the planet. Night by night the position of these points of light changed. They were clearly not stars, but moons. Galileo named them after his sponsors, the male members of the Medici family: nowadays they are called Io, Europa, Callisto and Ganymede – all lovers of the God Zeus, or Jupiter. Collectively, they are known as the Galilean satellites of Jupiter. Galileo's discovery was an important one, for it supported Copernicus' theory that the Sun and not Earth is at the centre of the solar system, for the four moons orbit around Jupiter just as Copernicus asserted that the planets orbit around the Sun.

Ever since Galileo's time, more powerful telescopes have progressively been developed. As astronomers have got a better and better view of Jupiter, more mysteries have accumulated around the planet and its moons. Jupiter has a banded appearance: the whole surface of the planet is decorated with belts and zones, apparently bands of different coloured clouds which lie along lines of latitude. In the southern hemisphere there is an enormous red spot, now about three times the size of Earth. The great red spot has been continuously observed for hundreds of years. Theorists speculated that it might be either an enormous

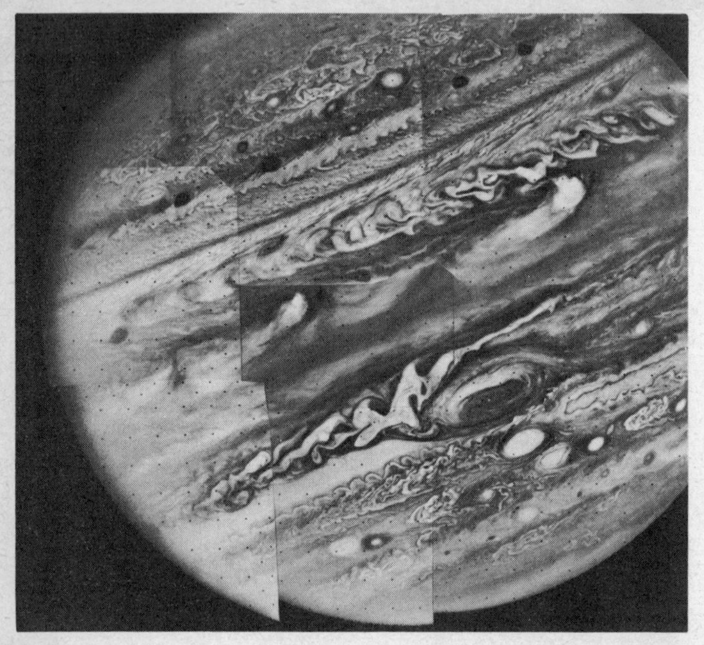

Photomosaic of Jupiter's atmosphere. The belts and zones run from lower left to upper right, and the southern hemisphere is dominated by the large, dark feature called the 'great red spot'.

cloud permanently hovering over a mountain on a solid core far below the visible surface; or a gigantic storm, like a hurricane on Earth, raging for hundreds of years as a result of the peculiar properties of Jupiter's atmosphere.

Nine more moons were discovered, but they are all much smaller than the Galilean satellites, mere chunks of rock probably much like the asteroids. Io and Europa, by contrast, are the size of the Earth's moon, while Ganymede and Gallisto are even larger, roughly the size of the planet Mercury. But other than a peculiar cloud of energised sodium atoms hovering around Io, not much more is discoverable through optical telescopes located on the surface of Earth. For the resolution of earthbound telescopes, or the capacity to distinguish detail, is limited by the turbulence of our own atmosphere.

However, the development of radio telescopes after the Second World War brought new surprises. Jupiter has an immensely strong magnetic field and is a powerful source of radio noise, in fact, this radio noise from Jupiter can sometimes be picked up on domestic radio sets. In some wavelengths, the radio noise arrives in bursts. These decametric radio bursts seem to depend on the relative positions of Jupiter's magnetic north pole and the moon Io. The theoretical astronomers Peter Goldreich and Donald Lynden-Bell suggested that Io generates about a million volts across its volume, and that the electrical circuit is completed in the upper atmosphere of Jupiter, powering a maser or microwave laser which sweeps invisible decametric radiation like a lighthouse beam into space. By the beginning of the 1970s, Io had been established as probably the most bizarre object in the solar system. Yet, owing to the limitations of earthbound observations, no detail had been observed on the surfaces of any of the Galilean satellites.

The Mission

During the 1960s planetary scientists working at NASA's Jet Propulsion Laboratory in Los Angeles made a discovery that was to bring long interplanetary journeys within the scope of rocketry. If a spacecraft passes close by a massive planet, the effect can be to sling the spacecraft outwards from the Sun. The huge energy needed to allow the spacecraft to resist the gravitational pull of the Sun and visit the most distant planets, is stolen from the motion of the massive planet that slings the spacecraft outwards. The planet itself is slowed in its motion by a minute amount, a few feet per trillion years, as its gravitational slingshot boosts the spacecraft on.

For just one month every 175 years, all the outer planets line up in such a way that a spacecraft from Earth can visit each of the planets of the outer solar system in turn. By the early 1970s unmanned robot spacecraft had already overflown Mars, Venus and Mercury. In the autumn of 1977 there was a rare opportunity to launch spacecraft to Jupiter, Saturn, Uranus, Neptune and Pluto in turn. This trip, called the Grand Tour, was fraught with uncertainties. No spacecraft, for example, had ever traversed the asteroid belt that lies between the orbits of Mars and Jupiter. It was also not known whether the electronics of a robot spacecraft could survive the radiation fields that were known to exist close to Jupiter. So first, ahead of time, in 1972 and 1973, NASA launched two relatively unsophisticated spacecraft to Jupiter as probes to see if the journey was possible.

Pioneer 10 and Pioneer 11 survived the voyage. Pioneer 10 was slung by Jupiter's gravitational slingshot into interplanetary space, becoming the first spacecraft from Earth to escape from the gravitational pull of the Sun. Pioneer 11 was programmed to travel on to Saturn before being, in turn, ejected from the solar system. The journeys of the Pioneer spacecraft proved that the asteroid belt was far more diffuse than had been feared and therefore safer for the passage of spacecraft. But the radiation belts surrounding Jupiter turned out to be more destructive than expected. Later spacecraft were to be armoured with radiation shielding as a consequence. The Pioneers also sent back pictures of Jupiter and its moons. The cameras aboard were fairly limited instruments, but even so Jupiter was revealed in more detail than ever before. There was a beautiful picture of the great red spot. For the first time the Galilean satellites appeared as more than mere points of light, but only just. Most important of all, the Pioneers set the stage for the Grand Tour of the solar system.

Originally, NASA planned to launch four identical robot spacecraft in the autumn of 1977. Two would travel to Jupiter, Saturn, and then on to Pluto. Two more would visit Jupiter, Saturn, Uranus and Neptune. However, the US Congress considered that the price was too high, so the plan was trimmed: there were to be only two spacecraft, and Pluto was to be excluded from mankind's first reconnaissance of the outer solar system. Voyager I was to travel to Jupiter and Saturn, Voyager 2 to Jupiter, Saturn, Uranus and Neptune. Voyager 1's itinerary was limited because the plan was to target it very close to Jupiter's moon Io, very close to Saturn's moon Titan, and then around the back of Saturn in such a way that a unique measurement could be made of the material which composes Saturn's rings. These constraints meant that after encountering Saturn, Voyager 1 would be hurled upwards out of the solar system by Saturn's gravitational slingshot.

But whatever the exciting possibilities, NASA had only thirty days within which to launch Voyager 1 and Voyager 2. Miss that launch window, and it would be necessary to wait until the year 2152 before a similar opportunity presented itself. In the event, all went well; within fourteen days, Cape Kennedy launched the two Voyager spacecraft.

The Spacecraft

The Voyager mission was intended from the start to be more a voyage of discovery than a strictly scientific exercise. The space-

Voyager spacecraft.

craft were to take a first close look at four planets and more than twenty moons. Virtually nothing, for example, was known about the moons they were to encounter. It was therefore essential not to prejudge the outcome by limiting the surprising observations the spacecraft might make. So the Voyagers carry on board a battery of instruments capable of registering an enormous range of phenomena. There are two imaging systems on each spacecraft; a narrow angle camera with a 1500 mm focal length, and a wide angle camera with a 200 mm focal length. These cameras take black and white still television pictures. By photographing the same scene three times through red, green and blue filters the spacecraft can take a colour picture, and by photographing the same scene a large number of times it can shoot a movie.

Visible light is only a small section of a much larger spectrum of electromagnetic radiation. Radio waves, for example, are the same phenomenon as light, but of a much larger wavelength – and radio waves are of course invisible to the human eye. The invisible light of shorter wavelengths than the visible is called ultraviolet, and that of longer wavelengths the infrared. The Voyager spacecraft were both equipped with instruments capable of detecting radiation in the ultraviolet and infrared. These instruments greatly extend the spacecraft's capacity to register surprising information. For example, clouds and haze are opaque to visible light but not to the infrared. So if the spacecraft were to encounter a planet or moon enveloped in clouds or haze, the cameras operating in the visible spectrum would be useless. But the infrared spectrometer would reveal information about what was happening underneath the cloud deck.

The spacecraft was also equipped with radio receivers. Listening in the same approximate range as a car radio, these can pick up radio waves associated with electrical discharges and allied phenomena in the environment of the outer planets. There are also other instruments tailored to gather information about the events going on in the huge radiation fields surrounding Jupiter and Saturn. The NASA scientists did not want to miss any of the surprises that come from looking at something clearly for the first time.

Because both Voyagers will eventually leave the solar system, they were each equipped with a gold disc which is a message to any alien civilisation that might eventually recover one of them. The face of the disc carries instructions on how to play the gramophone record inside, which carries the message. The message itself includes a collection of sounds, ranging from the sound of spoken Urdu to 'Johnny B Goode' by Chuck Berry. Also encoded on the record are 115 images of Earth. Britain is represented by a panorama of Oxford and a page from Newton's *System of the World.* Privately, NASA officials admit that it is extremely unlikely that any alien civilisation will gather up a Voyager spacecraft, but the gold disc is a reminder to the rest of the world that the Voyager mission is intended not merely as a flag-waving exercise for the United States, but as something done by humanity as a whole. Also, a message for aliens is something that the ordinary earthling may appreciate more than the results from an infrared spectrometer.

The heart of the Voyager spacecraft is a decagonal gantry that carries three large computers. These needed to be autonomous for signals took so long (45 minutes from Jupiter) to travel even at the speed of light from the spacecraft back to Earth that the Voyagers had to a certain extent to be capable of taking their own decisions. Attached to the back of the computers is a three metre diameter radio dish. This dish maintains communications between the spacecraft and Earth and transmits back the information. The dish transmits at the power of a twenty watt light bulb. NASA has three huge steerable radio antennae on Earth; at Goldstone in California, Madrid in Spain and Canberra in Australia. They are positioned around the globe so that, at any one time, one of them is always on the side of Earth facing the spacecraft. It is a tribute to the astonishing developments in communications that these dishes are capable of receiving television still pictures, composed of 64000 individual dots, at the rate of one every forty seconds, from a spacecraft antenna thousands of millions of miles away, radiating at no more power than a dim electric light bulb.

One could be excused for imagining that a Voyager spacecraft, equipped with three computers, a radio dish capable of transmitting information billions of miles back to Earth, two television cameras, instruments for detecting radiation in the ultraviolet, infrared and radio wavelengths, and a small nuclear power station, would be a large machine. Yet each of the Voyager spacecraft was fitted into a three metre wide capsule on top of the Titan-Centaur rockets which carried them into space.

The Planet

In March 1979 Voyager 1 encountered Jupiter after a journey of 500 million miles. Untouched by the asteroid belt, or by the belts of radiation that surround Jupiter, it successfully captured over 17000 images of the planet and its moons. And the pictures would have justified the trip by their beauty alone.

Jupiter, far from the Sun, is a twilight world. Streaking across its surface in great whorls and loops is a tremendous spectacle of colour – reds, oranges, yellows and whites. The movies shot by the spacecraft show this lurid world in rapid and violent motion, changing all the time, yet in overall pattern remaining strangely the same. Dominating Jupiter's appearance are the great belts and zones of clouds that mark the lines of latitude. Nestled between these streaks of coloured cloud are a multiplicity of spots, of all sizes and ranging in colour from red through brown to white. The spectacle is lavish and confusing, and provided the first great surprise of the Voyager mission. To understand the surprise it is necessary to know more about the planet Jupiter than what it merely looks like.

Jupiter is immense – many more than one thousand Earths would fit inside its volume – but it is not only size that makes Jupiter a quite different planet to Earth. Earth, like all the inner planets, is made of rock. Earth, Mars, Venus, but not Mercury, have thin atmospheres overlaying the rocky surface underneath. Jupiter and the other outer planets, by contrast, are almost all atmosphere. Jupiter's great size and its mass (which can be worked out from the periods of the orbits of Jupiter's moons) mean that Jupiter must be made almost entirely of hydrogen and helium gas. Much the same stuff, in fact, which astronomers think is the primordial composition of the Sun. Perhaps at the very centre of the sphere of gas is a small rocky planet, rather larger than Earth, but enveloped in an ocean of gas tens of thousands of miles deep. In the inner portion, the gas is so compressed by the overlying matter that it behaves like a liquid metal. All that Voyager saw of this planet were the tops of the clouds in the uppermost level of the atmosphere.

Jupiter radiates more energy than it receives from the Sun. This apparent impossibility can be resolved by considering that Jupiter is so large that the interior must still be warm and still cooling down from the heat of the formation of the planet when the solar system was formed.

The visible surface of Jupiter presented an immediate puzzle to the Voyager scientists, and in particular to members of the Imaging Team, who analyse the pictures sent back by the spacecraft. Different fluids moving relative to each other have an unerring tendency to mix. Since this must have been going on for millions of years, one could be forgiven for imagining that Jupiter would be a bland, monotonous planet. Yet, even though the different parts of the visible surface are in violent relative motion and in detail small features are constantly born and swallowed up again, overall the picture does not change. The very lurid, majestic pattern of all the different colours and shapes, the bands and the spots, remaining the same.

Scientists are rarely so lucky as to be presented all at once with a vast amount of new and detailed information about a hitherto obscure world. The members of the Imaging Team split into two schools of thought about why the surface of Jupiter looks and behaves as it does. One school suggested that everything could be explained on the basis of meteorology. In other words, that the patterns visible on Jupiter only extend downwards as far as the bottom of the upper atmosphere. The other school of thought maintained that the visible surface is a manifestation of processes going on deep within the denser liquid parts of the planet.

Garry Hunt is the only non-American in the Imaging Team. He is an Englishman who works most of the time at University College, London. His view that meteorology explained all had considerable merit not least because it was the simplest possible explanation and did not need to evoke some invisible process far in the interior of the planet. His view was also supported by a remarkable result from an American scientist outside the Voyager team, Gareth Wynn-Williams. Earth's atmosphere is distinctly divided into different regions by a phenomenon called the jet stream. In each hemisphere a cap of cold air sits over the pole and a belt of warm air over the tropics. These two regions of cold air and warm air are separated by a jet stream of rapidly flowing air. This jet stream is distorted by the influence of Earth's rotation into an irregular wave, which plunges the temperate regions alternately into hot tropical and cold polar air. This is the reason for the weather. At the poles or in the tropics there is no such constant alternation between different regions, only climate.

The Garry Hunt—Gareth Wynn-Williams thesis was that in Jupiter's atmosphere, many jet streams are stacked one on top of the other, giving rise to Jupiter's banded appearance. Meteorology is a poorly understood science, as incorrect weather forecasts constantly remind us, but computer programmes exist which fairly accurately model the overall features of Earth's weather. These programmes are very large and require the largest computers in the world to work themselves out. Gareth Wynn-Williams changed the numbers in a meteorological programme to suit the conditions appropriate for Jupiter. Although the calculation never settled down to a stable state, belts and zones and even spots appeared in much the same places as on the real Jupiter.

Andrew Ingersoll is an indigenous member of the Voyager team, in that he is also a member of the Jet Propulsion Laboratory's parent University, CalTech. He believed that Jupiter's upper atmosphere is a manifestation of processes going on deep below in the interior, and his view was supported by an experiment performed by Fritz Busse, from UCLA. Fritz Busse constructed a working model of Jupiter's interior. The interior, the liquid metallic core, he built as a solid sphere and heated it from within, much as is Jupiter itself by the residual heat of its formation. Surrounding this heated inner core was a shell of liquid water, representing the outer parts of Jupiter's interior. The whole model could be rotated rapidly – Jupiter, incidentally, rotates very fast for its size, about once every ten hours. When Fritz Busse got his model working, heated from within and spinning rapidly, columns appeared in the liquid going, as it were, north to south. Andrew Ingersoll suggested that this was an accurate model of Jupiter's interior and that the tops of these columns, which are in fact a series of nested cylinders, generate the belts and zones visible on the surface.

The truth turned out to be more interesting and puzzling than either of these two initial theories. A detailed analysis of pictures of the surface, taken at different times, showed that the belts and zones are not really relevant. What are constant are the winds, or jet streams, that lie along the latitudes. It appears that these jet streams are themselves powered by the spots (which are cyclones and anticyclones) lying between them. What powers the spots themselves is still a mystery, but there is a consensus now that perhaps the spots are caused to move by processes far inside Jupiter. As so often happens, the Voyager scientists were forced to reject their initial ideas, their processes of categorisation and their first intelligent questions, in order to pursue the deeper mysteries provided by the new information.

This new view of Jupiter's atmosphere throws into relief a long series of possible explanations for the most spectacular feature of the atmosphere, the great red spot. This feature, which has been observed continuously for over a hundred years, swirls in constant and violent motion. It is as large as four Earths put together, although there is some evidence that its size might have changed during recent history. Some astronomers had suggested that it is a huge hurricane, lasting for hundreds of years in the special circumstances of Jupiter's atmosphere; some that it is a cloud overlying a mountain or volcano far down in the interior, on the small rocky planet beneath; some that it is one of the exotic fluid phenomena that can be evolved in laboratory experiments, in fact like a solitary wave which can maintain itself almost indefinitely given the right circumstances. Unfortunately, nobody yet knows exactly what the red spot is, but they do know that it is none of the things that were suggested before Voyager sent back its wealth of images of Jupiter's surface.

As Voyager passed behind Jupiter, it photographed lightning flashes in the atmosphere on the dark side, and aurorae boreales. The radio antennae picked up transmissions produced by the lightning bolts, drawn into long whistles by their passage through space.

Both the Voyager spacecraft flew close to Jupiter, seeing it six months apart, but Voyager 1 and Voyager 2 flew close to different moons. Voyager 1's close encounter with Io produced the greatest surprise of the Jupiter missions. To get a proper perspective on what was discovered, it is better to view the planet-sized Galilean satellites not in the order in which they were encountered by the

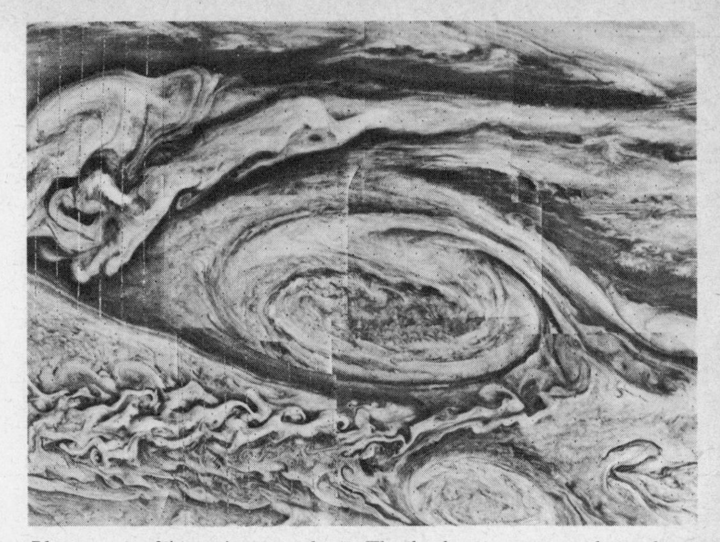

Photomosaic of Jupiter's great red spot. The clouds at its centre are lower than those on the outside.

spacecraft, but starting from the outside working in. Callisto, Ganymede, Europe and Io were the provinces of the geologists in the Imaging Team, a quite different set of people from the meteorologists. The encounters presented them with one of the greatest professional treats of their lives, for before Voyager, very little was known about those moons. In less than a week, during March 1979, these four planets swam into the ken of science.

Callisto

Callisto, like the other Galilean satellites, has no atmosphere. It is roughly the size of the planet Mercury and is thought to be made mostly of ice, with a small rocky core beneath. The Voyager images show a surface scarred by intense meteorite bombardment. The surface is saturated with impact craters, saturated because the addition of any more would only erase existing craters. In some places the energy of impact seems to have been enough to melt some of the ice to slush before re-freezing. One enormous impact appears to have shattered the surface, marking one whole side of Callisto with cracks.

Geologists use meteorite craters to estimate the age of planets.

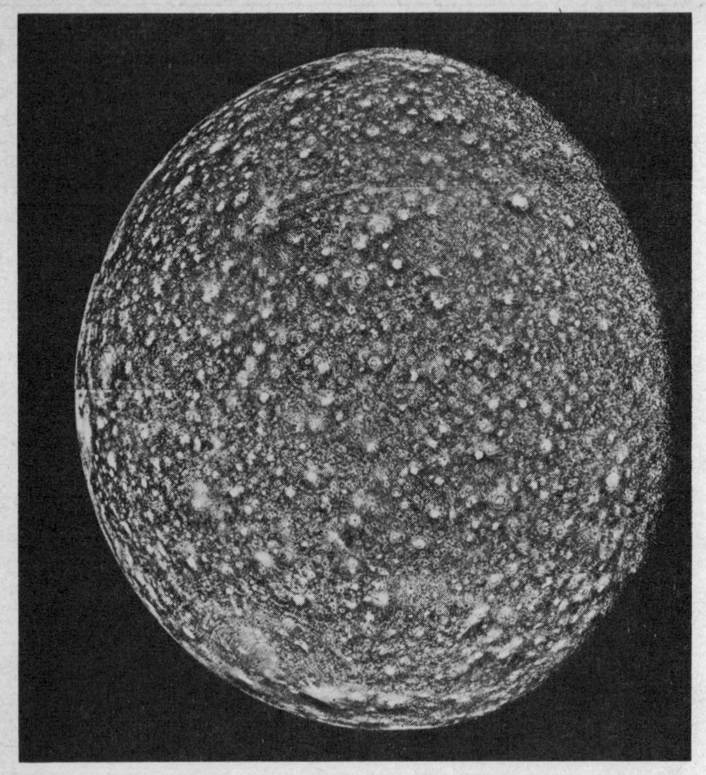

Computer-enhanced image of Callisto, outermost of Jupiter's large moons. The ice-surface is densely pitted with impact craters.

They suppose that the planets and moons of the solar system all formed from a huge disc of material swirling around the Sun, after the Sun itself formed from a collapsing cloud of gas. These events are thought to have taken place about 4500 million years ago. The date comes from studies of the radioactivity of meteorites (see Chapter 5), and from calculations of the age of the Sun. Exactly how and why the solar system formed from the huge disc of material is not yet known, but it is thought that soon after the Sun formed, the planets and moons generated themselves by accretion, gathering on to themselves the loose debris of the solar system by a process of meteorite bombardment. Although a few scattered fragments of debris still wait to be collected up in the solar system, the intense period of bombardment was relatively brief and the level of bombardment since has been comparatively negligible.

So Callisto, saturated by its icy surface with meteorite craters, has a very old surface. A surface to which nothing has happened since the accretion stage of the history of the solar system. The Voyager scientists realised that they were studying pictures of the oldest planetary surface yet discovered – a surviving record of the birth of the solar system.

Grooves on the surface of Ganymede, Jupiter's largest moon. The few visible craters provide a way of dating the surface.

Ganymede

Ganymede is the largest of the Galilean satellites, being slightly bigger than its outer neighbour Callisto, and again approximately the size of the planet Mercury. Like Callisto, the density of Ganymede, measured by determining both the exact size of the moon and the small gravitational pull on the Voyager spacecraft as it passed, reveals that it is composed of a thick mantle of ice overlaying a central rocky core. The surface of Ganymede is divided into two quite separate types of terrain. Parts of Ganymede look exactly like chunks of Callisto grafted on, for they are saturated with impact craters but, separating these sections of crater fields are regions striped by grooves, which make up about half of the moon's surface. The grooves, and there are many of them, run in parallel lines over the surface.

A similar groove exists on Earth according to one eminent American scientist, Eugene Shoemaker, former head of the Division of Planetry Sciences, United States Geological Survey. Shoemaker founded this division and himself selected the site of its headquarters, the small town of Flagstaff, Arizona, for it sits in the middle of one of the most varied places geologically in the United States. To the north of the town is an extinct volcano and scattered on the farther side are scores of cinder cones, relics of smaller eruptions. To the northwest the Coconino Plateau of Northern Arizona comes to an abrupt end at the Grand Canyon. To the south is the impressive Red Rock Canyon at Sidona; and to the east is probably the most perfect meteorite crater on Earth.

Meteor Crater is about a mile wide and 600 feet deep. It was formed about 25000 years ago when a meteorite no more than yards across crashed into Earth. If Earth's surface was not constantly subject to erosion and mountain-building, it would be covered in craters like Meteor Crater, as it is, it already shows signs of weathering into the landscape.

But more pertinent to the grooves on Ganymede, the country around Flagstaff also has a groove which runs across country for tens of miles, northeast of the town. Eugene Shoemaker believes it to have been formed by the plateau which it intersects stretching and pulling apart. To accommodate the pulling apart by increasing the surface, the groove dropped down below the level of the plateau. It seems reasonable to Shoemaker that the grooves of Ganymede were created in a similar way. Originally much like Callisto, Ganymede's surface was modified as the ice from which the moon is made expanded, creating the networks of grooves. The grooves themselves can be dated by counting the meteorite craters that lie on top of the grooved terrain, for these meteorites must have landed after the grooves were formed. The date is consistent with Shoemaker's explanation.

In one place on Ganymede, the grooves have slipped past each other. Perhaps this is the record of an earthquake fault. If it is, it is the only direct evidence of earthquakes anywhere else in the solar system beside Earth.

Moving inwards from Ganymede, the Galilean satellites change their nature. Europa and Io are smaller, about the size of Earth's Moon, and their densities are that of almost pure rock.

Europa: the surface is very smooth and shows a pattern of cracks similar to that seen on polar ice-caps on Earth.

Europa

Unlike Callisto and Ganymede, the surface of Europa has almost no meteorite craters. Instead what is visible appears to be a frozen ocean, rather deeper than Earth's. This ocean is marked with cracks that seem to have been infilled with slush, which subsequently refroze. As a result, Europa is the smoothest place yet discovered in the solar system, much smoother in scale than a billiard ball.

The near absence of impact craters means that the present surface of Europa was formed after the period of intense meteorite bombardment. What emerges from a comparison of Europa, Ganymede and Callisto is the standard history of a rocky planet or moon. The first stage is one of meteorite bombardment, which saturates the surface with craters. Then, part or all of this old cratered surface may be covered over or erased by other activity on the surface. Even this new surface becomes lightly cratered in turn by the few meteorites still lurking around the Jovian system. So perhaps the dramatic differences between Europa, Ganymede and Callisto can be resolved by discovering when their oceans froze. On Europa, the meteorites splashed down into a sea of liquid water, which only froze afterwards, so there are virtually no impact craters. On Ganymede, the ocean froze at about the same time as the bombardment stopped, erasing some of the original cratered surface with grooves. But on Callisto, the ocean froze before the end of the bombardment, making the surface a museum of the violent birth of the solar system. But why should Callisto's ocean have frozen before Ganymede's or Europa's? Jupiter itself must have been much hotter when it formed, so as the planet cooled, the oceans of the moons froze in turn starting with the furthest out.

This account, explaining the appearances of the three outer Galilean satellites, seems to make the process of rationalising quite new observation a relatively easy matter. But a perfect example of how puzzling an entirely new discovery can be was presented by Voyager 1's encounter with Io, the innermost of the Galilean satellites.

Io

Before Voyager, astronomers had guessed from their indistinct telescope observations that Io must be one of the most unusual places in the solar system. In the event, everyone was taken by surprise: Io is far stranger than anybody had imagined. It is roughly the same size as Earth's Moon, but there the resemblance ends, for Io's entire surface is blotched with bright reds, oranges and white. As Voyager 1 approached, the Imaging Team was expecting to see craters. There were craters, but not many, and they were the wrong shape. They appeared to be more like volcanic craters, which have irregular shapes and distinctively slumped sides. But if the craters were volcanic, and there were no meteorite craters at all, what had protected the surface of Io from meteorite impacts? On Io there is no sea to freeze over, for the moon is made entirely of rock.

Linda Morabito is a cheerful Italian-American in her late twenties. She works in the Optical Navigation Department at the Jet Propulsion Laboratory, in a quite different world from the isolated and exalted Imaging Team. Her job is to study the

This discovery frame provided the first evidence of vulcanism on Io. The diffuse circular cloud hanging over the top left of the moon was the first active volcano to be identified.

spacecraft pictures, looking for the positions of bright stars. This enables her to check if the spacecraft is straying off course. It is a relatively humdrum job, yet, to the initial chagrin of the members of the Imaging Team, Linda Morabito made the greatest discovery of the Jupiter encounters. She was using the graphics system on her computer to measure a picture taken by the spacecraft of Io with two bright stars in the background. She enhanced the contrast to reveal the stars and noticed a shape apparently hovering over the edge of Io. Morabito's first thought was that this 'anomalous crescent' was another satellite behind Io. She checked to find out if any of the other Galilean satellites were where her anomalous crescent was at that time – and they were not. It occurred to her that perhaps it might be a previously undiscovered satellite. A moment's thought convinced her that nothing of that size could have remained invisible from Earth. She contacted a camera expert to check if the shape was a fault in the imaging – and discovered that no known fault would produce such a shape. The only alternative left was that the anomalous crescent had something to do with the satellite Io itself. She discovered that the surface of Io underneath the crescent is marked by a large black heartshaped feature, something that had already been identified as a possible volcanic crater. But of course nobody had guessed that this volcano might be in the process of violent eruption. Linda Morabito herself came to the conclusion that the anomalous crescent was a volcanic plume, a cloud of dust and gas thrown up by the erupting volcano beneath. It was the first ever volcanic eruption to be witnessed on a body other than Earth in our solar system, and the discovery made Linda Morabito famous.

Almost at once another volcanic plume was discovered. Io appears to be in a state of continuous eruption. The colours of the surface are the colours of the various forms of sulphur, so apparently Io is covered in drifts and cones and lava flows of sulphur. Some of the erupting material escapes Io's gravity and settles into a wake of ionised gas behind the moon.

Just before the Voyager 1 spacecraft reached Io, a remarkable paper was published by three American scientists, Stanton Peale, Pat Cassen and Ray Reynolds. They proposed a mechanism which would explain Io's state of continuous volcanic eruption. Io is volcanic because it is sandwiched between the giant Jupiter and the other large satellites. Io raises a tide in the atmosphere of Jupiter in exactly the same way as our Moon raises a tide in Earth's ocean's, and the underlying mechanism and effects are entirely comparable. The gravity of Earth's moon distorts the seawater of the oceans into an oval. Because Earth is spinning, this oval of seawater is swept forward and always points slightly ahead of the Moon. The gravity of the oval of seawater in turn accelerates the Moon forward and pushes it out into space. In other words, the tides are gradually throwing the Moon out, away from Earth, and that is exactly what Jupiter's tides would like to do to Io. But the orbits of Io, Europa and Ganymede are locked together in a quite extraordinary way. From the standpoint of Europa, Io makes exactly two orbits for Ganymede's every one. The orbits are locked together and the effect of this on poor Io is to hold it firmly in place. Io cannot move, instead of being pushed out, it is alternately pulled and stretched by the immense gravity of Jupiter's tides. The heat of this constant deformation has melted the interior. A mere crust of sulphur and rock sits on a ball of molten lava, and that causes the volcanoes.

Voyager 2 was reprogrammed to shoot a movie of the volcanoes on Io. In the event it turned out to be something of a

Volcanic Io, seen against the background of Jupiter's turbulent atmosphere.

disappointment. Io can be seen rotating, but the volcanic plumes changed hardly at all during the time of shooting, and they appear merely fixed on to the moon like lumps of cotton wool.

Because of its constant eruptions, Io has literally turned itself inside out in the course of its history. Its surface, constantly sprayed with volcanic material, is the youngest ever observed. There are no visible meteor craters because they have been rapidly filled by the rain of volcano debris, and have disappeared.

Io, Europa, Ganymede and Callisto are all different and other unexpected finds kept cropping up during the encounters. Inside the orbit of Io, Voyager photographed the small moon Amalthea. Amalthea is potato-shaped, probably because its own tiny gravity is insufficient to pull its shape into a sphere. The asteroidlike moon is reddish in colour, and may have been painted by sulphur spewing out of Io's volcanoes. Jupiter's other small moons, outside the orbits of the Galilean satellites, are probably very similar to Amalthea, but neither of the two Voyagers flew close enough to see. However, the spacecraft did discover several very small moons that had never been seen from Earth and, most dramatically Voyager 1 discovered that Jupiter, like Saturn, has a ring. So much data came back from the two spacecraft encounters that getting the best out of the results may take ten years.

The results from Jupiter confirm that there is no place there where we can set down our spacecraft with any safety. The Galilean satellites, fascinating as they are, are all devoid of protective atmosphere and are bathed in lethal radiation generated in Jupiter's magnetic fields. All this is no comfort to those still looking for life, however microscopic. Even the gases in the atmosphere of Jupiter itself, which might have supported some kind of floating life, turn out to be moving too fast between very hot and very cold regions for there to be even a ghost of a chance.

In fact, the best hope for extraterrestrial life in the solar system has always been centred on the next staging post on our journey. Saturn is twice as far from Earth as Jupiter and slightly smaller. As a consequence, much less has been found out about it from Earthbased telescopes. Although Saturn itself is a gas planet like Jupiter, it does have an enormous moon, Titan, and this planet-sized object does have an atmosphere. Along with Saturn's astonishing rings, this was one of the prime targets of the two hard-worked Voyager spacecraft. What they found out had less to do with the search for life in the solar system and much more to do with the origins of the solar system itself.

4 Resolution on Saturn

Fisher Dilke

What was known before Voyager 1

In November 1980, NASA's unmanned robot spacecraft Voyager 1 passed close to Saturn, its rings and some of its many moons. The pictures and data transmitted to Mission Control at the Jet Propulsion Laboratory in Pasadena caused a sensation, a sensation because no human being had ever seen Saturn close up before, and because the Saturnian system proved to be very different from what had been expected. Even now, many of the discoveries have not been resolved. Voyager 1, did not, of course, make the initial discovery of Saturn; that event happened far back in antiquity, for Saturn is a very bright object in the night sky. What Voyager 1 did was to improve the resolution on Saturn by about ten thousand times. (Resolution is an important word in astronomy, being a measure of the clarity or accuracy of an observation. To be precise, it is the distance two parallel lines on the observed object need to be apart from each other for the observer to notice that they are two, and not one line.)

In 1610 Galileo looked at Saturn through his telescope. It gave him the advantage of much greater resolution; everyone who had looked at Saturn before had done so with their unaided eyes. As is often the case, greater resolution presented a puzzle: the planet appeared to have large ears. Galileo never explained the ears, for his telescope was not powerful enough. The problem was solved later in the seventeenth century by the Dutch astronomer Christian Huygens: Saturn has not ears, but a magnificent set of rings.

The fundamental problem of planetary rings was not solved until the nineteenth century, when the great Scottish physicist, James Clerk Maxwell, worked out the dynamics of Saturn's rings. Maxwell's theory depends on the theory of orbits developed long before by Isaac Newton. Newton's theory of orbits can be explained as follows: imagine that one drops a cannonball off a tower exactly sixteen feet high. The cannonball will fall to the ground in one second. Now imagine that the ball is pushed horizontally off the tower and given a horizontal velocity of 100 feet per second. The ball will still fall to the ground in one second, but in this time it will also travel 100 feet horizontally. Now imagine that the ball is given an enormous sideways push off the tower, so much that, in the first second, it travels so far horizontally that the Earth (for Earth is of course spherical) drops away exactly 16 feet. The ball will still have its horizontal velocity, and will still be 16 feet above the ground. It will not hit the ground, unless there is a mountain in the way, for it will be in orbit, in a state of constant free fall. It is a consequence of the theory of orbits that inner orbits are faster than outer orbits and this made an understanding of Saturn's rings very hard to come by.

If Saturn is observed over many years, the plan of the rings changes in relation to Earth. When the rings are edge-on, they are invisible. This means that the rings must be less than a few miles thick. James Clerk Maxwell discovered that the rings cannot be a solid sheet of material, for no known substance of the dimensions of Saturn's rings would be able to stay rigid under the shear of the orbits of the outer parts versus the orbits of the inner parts. In other words, Saturn's rings must be composed of a myriad swarm of small independent moonlets, less than a few miles in diameter at most, and probably only pebble-sized.

As telescopes improved in resolution, it became clear that the ring boulders are not evenly distributed. In fact there are clear patches, and parts of the rings that are more diffuse than others. By this century, telescopes had become advanced enough for astronomers to get some idea of conditions not only in the rings of Saturn, but also on its moons orbiting outside its rings. According to standard astronomical practice, the moons of Saturn are named after the mythological associates of the god who is called Saturn in Latin, and Cronos in Greek. So the moons are called Mimas, Enceladus, Tethys, Dione, Rhea, Titan, Hyperion, Iapetus and Phoebe. Of all these moons, by far and away the largest is Titan, which was thought to be the largest moon in the entire solar system.

In 1944 G. P. Kuiper discovered that Titan has a substantial atmosphere, including at least some methane, a component of natural gas on Earth. This makes Titan unique among the moons of the solar system in possessing a true atmosphere. By the 1970s the observation had sunk into the astronomical community, and there was a speculation that a greenhouse effect in Titan's atmosphere might make conditions warm enough there for extraterrestrial life to exist on the surface.

In April 1973 the unmanned probe Pioneer 11 was launched to Jupiter and Saturn. Its Saturn encounter was followed by a fairly close Titan encounter. There was a great deal of excitement about this Titan encounter, for the few pictures sent back from Pioneer 11 might resolve some of the mysteries of Titan. For example, Titan was known to have red clouds: are there gaps in the clouds through which the surface can be seen from space, as is the case with Earth?

The United States and the Soviet Union collaborate with each other to a certain extent on deep space missions. During Pioneer 11's encounter with Saturn and Titan, the USSR agreed to restrict reconnaissance overflying and interference near the three large radio dishes, spread around the globe, that collect the information sent back from the USA's unmanned probes in deep space. But it seems that a clerk in NASA made a mistake in the timing of the Saturn – Titan encounter. Innocently believing that the encounter was over, a Russian reconnaissance vehicle jammed the information coming in. So no pictures were recovered of Titan from Pionner 11. The *National Enquirer*, a sensational paper based in Florida, published a story that pictures *had* come back, showing an alien base on Titan, but that the information had been suppressed by the American Government, with an appropriate cover story.

Unfortunately very little is discernible through Earthbound telescopes of Saturn's other large satellites. Before Voyager, there was an informed guess that many of them must be made of ice. So the Saturn section of the Voyager 1 mission was planned around what were thought to be the two most interesting objects in the Saturnian system: Titan, which might harbour life on its surface, and the rings.

The Mission Plan

The trajectory engineers and mission scientists at the Jet Propulsion Laboratory had two spacecraft at their disposal, Voyager 1 and Voyager 2. The trajectory for Voyager 1 was determined by three targets: the close encounter with Jupiter's moon Io, which unexpectedly turned out to be volcanic; a close encounter with Titan, and a very special observation of Saturn's rings. Everything else was secondary, and in fact to achieve these three

Single frame from a computer animation of Voyager 1's encounter with Saturn. The sun is at upper right, earth to its immediate left.

targets Voyager 1 itself needed to be sacrificed, for the trajectory after Saturn would take the spacecraft up and out of the plane in which all the planets orbit the Sun, and into interstellar space.

At Saturn, it was decided to encounter Titan first. There was a possibility that Voyager 1's close pass by the rings might damage the spacecraft by collision with a ring boulder. If this happened, a Titan encounter by a defective spacecraft would prove as frustrating as Pioneer 11's abortive Titan encounter. Titan orbits Saturn every sixteen days, so every sixteen days there was a time window through which Voyager 1 could travel on to its next major objective, a radio observation of the ring boulders. Passing on one side, but below Saturn, the spacecraft would then be thrown by gravity up behind Saturn and would shine two radio beams back to Earth through each part of the rings in turn. This information would enable the scientists of the Jet Propulsion Laboratory Radio Science team to determine the sizes of the boulders that make up the rings in all of their parts. As many other targets as possible were fitted in to the basic pattern. Close views were planned of the planet Saturn itself, and the moons Rhea, Dione and Mimas. Voyager 2, following one year behind, would have closer encounters with Tethys, Enceladus, Iapetus, and Hyperion and with the rings.

Composite photograph showing Saturn and its moons. Anticlockwise from the top: Rhea, Enceladus, Dione, Tethys, Mimas and Titan.

The Planet

Saturn is the second largest planet in the solar system, and is distinctly flattened by its rate of rotation. The immense planet, into which more than 650 Earths would fit, rotates once every ten hours and forty minutes, about the same as Jupiter. It is shrouded in yellow haze, through which are visible a few details of a lower level in the atmosphere that resemble the lurid and spectacular surface of Jupiter. Underneath is an ocean of liquid hydrogen more than 30 000 miles deep, at the centre of which is a rocky core, rather larger than Earth. Saturn being mostly made of gas, is a much less dense object than Earth: it is so light that if one could only find a bucket of water big enough, the planet would float.

One of the scientific problems associated with Saturn is that it behaves rather like a star. It gives out more energy than it receives from the Sun. The magnitude of this internal energy source is very large and difficult to explain away. Jupiter has an internal energy source, but this can be understood on the basis that the planet's interior is still hot from its formation and that the internal energy source is this primordial heat slowly leaking away. The same process must be happening in the interior of

Saturn, but the planet is leaking away too much energy for this to be the whole story. (Perhaps the discrepancy is caused by a redistribution of material within the planet, the heavy substances raining downwards, the lighter welling up.)

Saturn's surface is rather disappointing, the yellow haze all but obscures the interesting cloud shapes underneath. The Jet Propulsion Laboratory, though, has spent years in developing the new technology of image processing. Each picture that comes back from the Voyager spacecraft is a matrix of 64 000 picture elements or pixels. In other words, they are television pictures rather than photographs. Each pixel has 256 digital levels of grey. A Voyager image is essentially much more than just one picture, for by selecting different ranges in the grey levels, many pictures can be manufactured from any one image. Confronted by a hazy scene, with just a few features peeping through, the image processors can manipulate the image to bring out such detail as exists. Performed by high-powered computers, the process is roughly analogous to developing a photograph.

A planet very similar to Jupiter emerged from underneath the haze. Like Jupiter, Saturn has a red spot, and long swirling vortices that trail over its surface. Underneath the haze, it appears to have all the lurid colours that make Jupiter's surface so exciting. But aside from its hazy yellow beauty the planet itself did not produce the surprises which became almost commonplace elsewhere in the Saturnian system.

Titan encounter

Voyager 1 sent back a new picture, or its equivalent, every forty seconds. These pictures appeared on television screens scattered around the Jet Propulsion Laboratory. Occasionally the screens drew crowds of people pushing for a first look at some moon or a part of the rings. But before the encounter, the screens mostly showed pictures of Saturn. As day followed day and hour followed hour, the planet increased in apparent size until the pictures no longer showed all Saturn and its rings, only a cropped part. Then, on 10 November, a new character appeared, swimming towards the spacecraft for its carefully planned close encounter – Titan. Unlike Pioneer 11's mission, nothing went wrong with the transmission of Voyager 1's pictures, even though the spacecraft was nearly a billion miles from Earth, travelling at about 30 000 miles an hour, and the pictures were taking an hour and a half to return.

The watchers at the screens were eagerly hoping for a glimpse of the Titanian surface through a gap in the clouds. Closest approach happened on 11 November. It was disappointing, for Titan presents an almost featureless face. The moon is a fuzzy orange ball. The only features visible are a north polar cap made of high altitude haze which is rather darker than Titan as a whole, and a colour difference between the northern and southern hemisphere; the northern hemisphere is a darker shade of orange. But the general impression is of looking at a moon enveloped in a new uniform layer of orange haze or fog.

Titan might still be a mystery if the people who had planned the Voyager mission had not had the foresight to equip the spacecraft with an infrared spectrometer. The IR spectrometer looks at the infrared radiation, or heat, coming from a planet or moon. Infrared radiation is much more penetrative of cloud and haze than ordinary light, but the IR spectrometer aboard Voyager 1 was not programmed to take a standard 'picture' of Titan. What it produced was an infrared signature of the atmosphere at all its levels, for chemicals present at different levels emit quite characteristic call-signs in the infrared. So what the infrared experiment promised to provide was an inventory of the chemicals present in the Titanian atmosphere, as well as a location of the surface and a surface temperature.

The information that came back from the spacecraft, and then emerged from the many computer calculations done on it, looked at first sight rather unpromising and unglamorous – a mere graph. Yet the various sections of the graph give clear evidence of the existence in the Titanian atmosphere of the carbon based gases methane, ethylene and hydrogen cyanide, amongst others. The graph also showed that the surface of Titan might be partially covered by an ocean of liquid methane, a component of natural gas. But the most sensational discovery was the hydrogen cyanide, for although this chemical is highly poisonous, it is thought to have been a crucial chemical for the evolution of life on Earth. Its presence on Titan does not necessarily mean that there is life there, for it is probably far too cold – about 180 degrees centigrade below zero – but it indicates that, given the right conditions, life might well start there.

What is it like on Titan? High up in the atmosphere, radiation trapped in Saturn's enormous magnetic field is creating a haze of tiny particles, almost certainly organic compounds. These particles, like manna from heaven, drift down through the orange clouds, through an atmosphere made mostly of methane and nitrogen. The manna falls in soft drifts on the surface, or into the sea of liquid methane. And all the time, the atmospheric chemistry is manufacturing the components of life.

In 1953 the American chemist Stanley Miller did a brilliant and obvious experiment that nobody had had the wit to do before. Into a closed flask he put water like that of the sea before life, and gases in the right proportions of the early atmosphere. He made sure that the interior of the flask was sterile and then set off a continuously flashing spark inside it. The effect of the spark was intended to be similar to the effect of lightning in the atmosphere of the pre-biotic Earth. Within hours a brown sludge began to form in the water. The sludge contained amino acids, the building blocks of proteins, and nucleic acids, the building blocks of DNA, the chemical of inheritance. Miller's experiment suggests that, given the right conditions, life happens automatically on any planet. The same experiment has been done on a simulation of Titan's atmosphere and the result is the same, with this difference. It seems that Titan is too cold for the process fully to take place; but warm Titan up and life might start.

So Titan, perhaps, looks much as the Earth did before life began. Although it is now in deep-freeze, its moment may come. In about 5000 million years' time the Sun will run out of the hydrogen fuel that currently makes it shine. It will then become a red giant and, expanding enormously, it will engulf Earth. Then Titan will warm up and, for just a few million years until the Sun contracts back to a tiny, vestigial, white dwarf, may have the right conditions for life. If we have any descendants then, Titan may provide a staging post for them on the way to finding a new planet and a new star after the catastrophic loss of Earth.

As a postscript to the Titanian discoveries, the Voyager scientists discovered that the atmosphere is so deep (much deeper than Earth's) that Titan is not after all the largest moon in the solar system. That honour belongs to Jupiter's moon Ganymede. On the other hand, no NASA official wants to send a lander to Ganymede. After the Voyager mission, Titan is top of the list for a lander mission, even if the lander has to be in the form of a boat. We know what Ganymede looks like but the surface of Titan really defies the imagination. We need to send a robot spacecraft there to take a look.

Rhea

When a spacecraft flies by a moon, its trajectory is altered slightly by the gravity of that moon. From this slight alteration in course the mass of the moon can be worked out. The volume of the moon can be determined easily by measuring photographs that the spacecraft takes of it, knowing how far away from the moon the spacecraft was when it took the photographs. Combining the two, the mass and the volume, gives the density. Except for Titan, the density of the moons of Saturn is very close to, in fact slightly above, that of ice. So it is reasonable to conclude that the Saturnian satellites are made of ice mixed with a little rock.

Beside Titan, the moon that Voyager 1 took the closest look at was Rhea. Rhea is just under a thousand miles in diameter. It is an intensely cratered object, rather like the planet Mercury, but smaller. Like the other moons, it appears to be made of ice, with a small admixture of rock.

Almost immediately the pictures came in from the spacecraft, the surface of Rhea presented the Voyager geologists with a problem. Rhea is covered with large and small impact craters, but in one segment of Rhea there are no large impact craters at all. That area is so extensive that it is not feasible that the absence of large craters happened by chance.

To understand the enigma, it is necessary to know the way in which the solar system is thought to have formed, about 4500 million years ago. It is thought that the Sun and planets formed from an immense cloud of material that collapsed in on itself owing to the mutual gravitational attraction of all its constituent

Photomosaic of a part of Rhea. Very few large craters can be seen to the right of the centre line.

parts. The same phenomenon can be observed elsewhere in our galaxy, for example in the Orion nebula, where there are dark spots called Bok Globules. These are thought to be stellar systems observed in the process of formation, before the central star begins to shine. In the first stage of formation of the solar system, the cloud of material collapsed, and the central part became the Sun, which began to shine. The rest settled into an enormous disc of material swirling around. It was from this disc that the planets and moons formed by a process called accretion, that is, accumulation by the impact of small bodies on to their surfaces. In other words, shortly after the birth of the solar system there was a period of intense meteorite bombardment which made the planets and moons. In the case of Earth the evidence of this bombardment has been erased by geological processes and erosion. Saturn's moon Rhea, however, has preserved on its surface the characteristic jumble of meteorite craters standing shoulder to shoulder, evidence of the process which formed it.

So how is it possible to explain the area on Rhea that is unmarked by large craters? The most plausible idea is that, during the formation of the satellites of Saturn, there were two distinct phases of bombardment. The first included both large and small objects. Then something happened to Rhea: part of it was resurfaced, perhaps by icy slush coming up from inside to erase the craters over a large area. Then the second phase of bombardment began. This time there were only small objects, making small craters. So the resurfaced area was marked only by small craters. It is a plausible idea, because it is possible to imagine one whole class of objects circling the Sun and battering Rhea, and another and probably different class of objects circling Saturn and battering Rhea, nor would the times of battering by the two sets of objects necessarily have been the same.

The Inner Icy Moons

The next moon in from Rhea is Dione. Like Rhea, Dione is largely made of ice, but it is smaller, about 700 miles in diameter. The surface of Dione is patterned with areas of coloration, called wispy terrain, a description coined by one of the Voyager Imaging Team geologists, Lawrence Soderblom, which has stuck ever since. The best explanation for the wispy terrain is that at some stage the surface cracked and material, perhaps gases, emerged from the inside. Parts of the icy cratered surface were then painted with wisps. Dione has a minute companion, a tiny moon called Dione B. Dione B orbits Saturn in exactly the same orbit as Dione, but ahead. So Dione never catches up Dione B, and the two will probably continue chasing each other around until some external agency intervenes. Clearly visible in the Voyager images of Dione are great cracks or chasms running across the surface. Dione shares this property with two of its sisters, Tethys and Mimas.

Tethys is only slightly smaller than Dione, about 650 miles in diameter. Its density is almost exactly that of ice, so there can be very little rock in Tethys. One side seems to be marked with an enormous impact crater and the other by a gigantic chasm, larger than the Grand Canyon. The chasm is so large in relation to Tethys that it is reasonable to suppose that it is the visible part of a crack running right through the volume of the moon. Ice, by far and away the main constituent of Tethys, behaves quite differently in Earthlike temperatures from Tethyslike temperatures. At the temperature of Tethys (roughly 200 degrees centigrade below zero), ice is very hard and brittle, so it cracks easily if it is struck. Ice at room temperature is, in comparison, a slushy malleable substance. So it seems plausible to suggest that a large meteorite crashed into Tethys, made the enormous crater, and cracked Tethys in two. In fact, Tethys and Dione are probably large spherical piles of icy rubble held together by their mutual gravitational attraction. Yet Voyager 2 discovered that Tethys, like Dione, has small companion moons occupying the same orbit. Tethys has two. They are small, irregular shaped objects which look rather like potatoes.

The next moon in Enceladus, is quite different. The best pictures of Enceladus came from Voyager 2; they show a surface that is cratered in some places and in other places seemingly resurfaced with networks of grooves that look very similar to the grooves on Jupiter's moon Ganymede. These resurfaced areas appear to be quite recently formed, indicating that Enceladus is still geologically active. Like Jupiter's moons Io, Europa and Ganymede, Enceladus is held in a resonance orbit. Perhaps internal heating, caused by Saturn's tides, has warmed Enceladus enough to melt the interior and maintain geological activity on the moon, allowing resurfacing to take place periodically. Whatever the underlying reasons, Enceladus has a dramatically different appearance from its neighbours.

Mimas, next in from Enceladus, resembles Tethys and Dione more than its neighbour. About 250 miles in diameter, Mimas is

Mimas.

cratered and cracked, one side being marked by an impact crater so enormous that it seems inconceivable that Mimas should have taken such a knock and survived. The crater dominates the entire side of the moon, giving it a weird appearance rather like an eye. Meteorites impacting a planet or moon travel very fast – about six miles per second in the vicinity of Saturn – and experiments suggest that there is a smallest size of moon that can withstand large meteorite impacts. Below this critical size the moon is either broken apart into many pieces which then collect up again or it is broken into a few different pieces which stay separate. Tethys and Mimas are probably rubble piles, and the two moons that orbit Saturn inside the orbit of Mimas, S-10 and S-11, may be the remnants of a once larger moon that was split in two by a meteorite impact.

S-10 and S-11 have intriguing properties. They are so called because the International Astronomical Union has not yet given them classical names, they were merely the tenth and eleventh satellites of Saturn to be discovered. S-10 is roughly 120 miles in diameter, S-11 slightly smaller, about 80 miles in diameter; both moons are very irregular in shape. Although they look as if they would fit together into a rough sphere, this is not the only evidence that they may once have been the two halves of a larger moon. S-10 and S-11 move around Saturn in almost exactly the same orbit, but because it is not quite the same, the one in the inner orbit, moving faster, is slowly catching the other one up. Every four years they ought to collide, but it is thought that as they approach one another, the gravitational influence of each on the other causes them to exchange orbits. The faster one is slowed into an outer orbit, and the slow one is accelerated into an inner orbit. It now starts the long journey of catching the other up, with the distance it has to make up being the exact circumference of its orbit around Saturn. S-10 and S-11 meet every four years, when their little dance is repeated as it must have been for millions if not billions of years.

The Rings

The Rings of Saturn are one of the great sights of the solar system. Thirty-eight thousand miles wide, and less than a few miles thick, they are composed of a gigantic collection of boulders of ice. The rings are not uniform in appearance, the main body of the rings being divided in two by an apparently empty band called the Cassini Division, which is 2188 miles across. The main part of the ring outside the Cassini Division is called the A-ring. Even seen from Earth, the A-ring is divided, by Encke's Division. The part of the main ring system inside the Cassini Division is called the B-ring. Inside the B-ring is a much more diffuse area of the rings, called the crepe or C-ring. Even further in towards Saturn is a D-ring. The diffuse material stretching far out into space beyond the A-ring is called the E-ring or extended ring. And just off the edge of the A-ring is a narrow ring, like a line, which is called the F-ring. The nomenclature seems absurdly complicated, but even so it is quite incapable of describing the sheer complexity of the rings that was revealed by Voyager 1's cameras. The rings are full of structure, everywhere you look there are divisions. In fact, the suspicion is growing that the closer you look at the rings, and the more resolution you bring to bear on them, the more structure you will find. The most surprising thing that Voyager 1 discovered, though, was that there are dynamic changes taking place in the rings – in other words, they are not static, things are happening to them right now.

The Rings of Saturn: computer-enhanced image showing the complexity of their structure.

The rings lie within what is called the Roche Limit of Saturn. This is the distance from Saturn within which any moon will be broken apart by the close gravitational influence of the planet. So the rings may represent an icy moon or two which were never able to form. As such, the rings are a remnant of that time, about 4500 million years ago, when Saturn's moons formed from a much larger ring which stretched out beyond Titan to Iapetus and Phoebe. All the outer parts solidified into moons through a process of accretion: the inner parts were unable to do so and have remained in their primevial form for billions of years. So, looking at the rings speeded up in time, one would see a spectacle much like an enormous carousel, with the inner boulders of ice speeding around faster than the outer ones. The carousel has been turning for so long – 4500 million years – that it seems reasonable to suppose that anything that was going to happen would have happened long ago. However as Voyager 1 approached Saturn, its cameras picked up curious dark markings in the B-ring. These appeared to change with time, so a sequence of pictures was taken which was made into a movie which showed that the B-ring has spokes radiating out from the

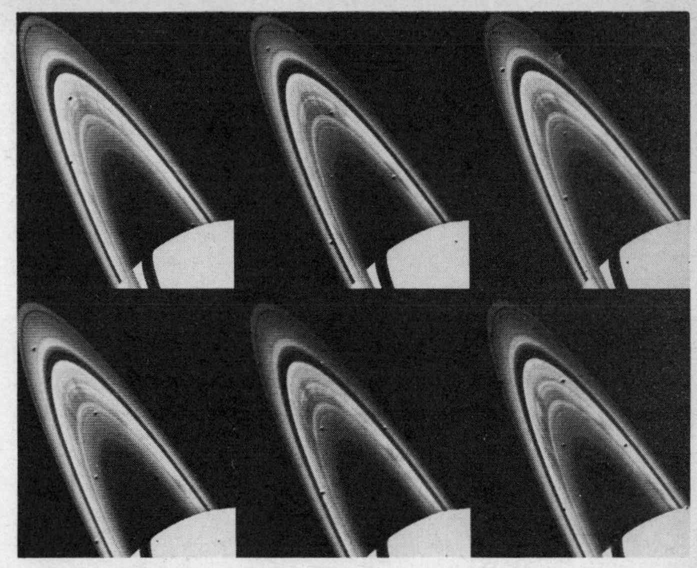

Six frames from the film of the Voyager 1 Saturn approach which first showed the existence of spokes in Saturn's B-ring.

central planet which travel around the ring like a succession of dark bands. The spokes are only visible on one side of the rings and appear to be generated in the shadow behind Saturn. From the point of view of Voyager 1's approach, 'spoke' seemed an appropriate name, but the view was very oblique, and in general exact shapes are difficult to make out from oblique angles. So the spacecraft was reprogrammed to shoot a movie of the spokes as it left Saturn.

Voyager 1's trajectory carried it above Saturn as it left, so it could look down at the spokes instead of glancing sideways at them. The new movie showed that the spokes are very irregular in shape. But the most remarkable difference in this second movie was that the spokes had changed colour. Approaching Saturn, with the Sun behind the spacecraft, the spokes appeared dark. Leaving Saturn, with the spacecraft looking back in the general direction of the Sun, the spokes appeared light. This was a very important clue as to what the spokes actually are, for the property of changing colour with the direction of view is shown only by very small objects. Dust viewed with the light coming from behind one's head appears black, whereas seen from the other side, backlit, it appears light. The spokes must be made of very fine matter, probably dust or a frost of ice. The current theory of the spokes then is that some mechanism operating in the Sun's shadow behind Saturn levitates clouds of dust or ice above the rings. These clouds orbit with the rings and are then swept out by Saturn's magnetic field, which probably contributes to their shape. It is extraordinary to consider what it must look like close up: clouds of frost or dust whirling over the orderly procession of orbiting rings of boulders.

One of the principal objectives of the mission was to discover the sizes of the individual ring boulders. This objective was considered so important that the Voyager 1 spacecraft was effectively sacrificed to achieve it. Now came the time when Voyager 1's trajectory was to carry it below Saturn and up the other side, in such a way that radio beams from the spacecraft would shine back at Earth cleanly through each part of the rings in turn. Radio waves are made of the same stuff as light. If you shine a laser beam through different kinds of cloth, and look at the laser beam through the cloth, the patterns you see are determined by the nature of the cloth, so that it is possible to distinguish between cloth made from thick threads and cloth made from thin threads. It is also possible to distinguish tightly woven cloth from loosely woven cloth. Thus Voyager 1's task of shining its radio beams through the rings was expected to reveal the size of the boulders of ice that compose the rings as well as information about how closely packed they are together.

The results, which took months of calculation to turn into simple information, were rather surprising. The boulders in the innermost large ring, the C-ring, are on average two metres in diameter. In the central B-ring, which has the spokes, the boulders appear to be so packed together that the radio signals were unable to penetrate. In the outer A-ring, the boulders are on average ten metres across. Before Voyager, the boulders were thought to be pebble-sized, but it now seems that they vary between approximately the size of a car and the size of a large house. Of course, these are only average sizes, so some must be the size of pebbles and others the size of office blocks.

Because they are nearer to Saturn, the boulders of the rings orbit faster than the moons. Imagine a part of the rings where all the boulders are orbiting at exactly twice the rate at which the moon Mimas goes around. Exactly every second orbit, Mimas will line up with each one of these ring boulders. Over millions of years, the accumulated effect of Mimas' gravity will pull these ring boulders out of circular orbits and into elliptical orbits. Having elliptical orbits will cause the boulders to collide with their neighbours, slow down and move out. So, in those parts of the rings where the boulders orbit at twice, or for that matter three or four times the orbit of Mimas (or any of the other inner moons), one would expect parts of the rings to have been sucked clean of boulders by the gravitational influence of the moons. It is called the resonance theory because it predicts that there should be clear zones between the rings at the resonant frequencies of the moons' orbits. It is a very satisfying idea and everyone was very fond of it but, when Voyager 1 took a close look at the rings, the resonance theory regretfully had to be abandoned. The theory, for some reason, is wrong; it does not fit the facts. Not only are the real gaps in the wrong places, there are too many of them. This left the Voyager scientists with a problem: if the resonance theory is wrong, what is causing all the structure in the rings?

A clue came from the F-ring, which is a threadlike circular structure outside the main body of the rings, for, far from being an even circle, it turned out to be braided and kinked. In one place, there appear to be two F-rings, twisted around each other. How can streams of icy boulders twist around each other in a double helix in conventional gravitational theory? A not entirely obvious but puzzling feature of the F-ring is its very existence, for under normal circumstances it should have dissipated sideways, looking more like a diffuse sheet than a line. Yet the F-ring is very narrow, so some process or processes must be confining the boulders in place.

This might be the two small moons which orbit very near the F-ring, S-14 just inside and S-13 just outside. The inner moon, S-14, orbits Saturn slightly faster than the boulders of the F-ring and, as it passes, it slows the boulders and pushes them into an outer orbit, confining the inner edge of the ring. The outer moon S-13 orbits slower. As the F-ring passes S-13 its gravity pushes the ring in on the outer edge. S-13 and S-14 behave like shepherds, fencing the F-ring into a narrow line.

The shepherd moons might also cause the braiding. The gravitational effect that allows the moons to confine the ring on the inner and outer edge acts more strongly on more massive boulders than on small ones. These more massive boulders are not gently confined, but pushed away, carrying streams of other boulders with them whenever a shepherd moon passes. This effect might cause the unusual shape of the F-ring.

The phenomenon of shepherd moons is not easy to understand. Gravity is an attractive force, not a repulsive one, so how can a shepherd moon confine the edge of a ring? Straightforwardly, it is reasonable to imagine that it would do the opposite. The difficulty is partly resolved by realising that there are three gravitational objects involved: the individual ring boulder, the shepherd moon and Saturn itself. Gravity involving three bodies can produce apparently contradictory effects between any two of them, including the shepherd moon effect.

Great excitement was generated by the discovery of the F-ring shepherds, for the many divisions in the main body of Saturn's rings might after all be caused by small shepherd moons embedded in the structure of the rings. It is an attractive idea, and it fills the gap left by the demise of the resonance theory. Unfortunately, none of these shepherd moons has been found, even though Voyager 2 was reprogrammed especially to search for them.

Although the resonance theory did not explain the positions of the gaps in the rings, some of the Voyager scientists started to find it appropriate to other features in the rings. In certain places in the A-ring, the material seems to be collecting together. Some process appears to be gathering the ring boulders into narrow rings which are so crowded with material that they do not let the sunlight through. These points of concentration are at resonance locations with Saturn's moons. This could be a very important discovery, for it may be a clue about how Earth, Saturn and indeed all the other planets may have formed, over 4000 million years ago. For it is not known exactly how the primeval disc, much like Saturn's rings only enormously larger, became the planets. Why, for instance, should the planets have formed in the exact locations in which they did? It is extremely fortunate for us, for example, that Earth is the distance from the Sun that it is, for if it were a few million miles closer it would be too hot for water to exist as a liquid. If it were a few million miles further out, it would be too cold for water to exist as a liquid. Liquid water is almost certainly essential for life. Neither hotter Venus nor colder Mars sustain life. So what made the portion of the primeval ring in the neighbourhood of the orbit of Earth condense into a planet just here?

A clue is provided by the resonance locations in Saturn's rings. There, the material of the rings is coagulating, and the cause of this coagulation appears to be the resonance influence of Saturn's moons. How is this relevant to the formation of the planets? Perhaps the massive outer planets Jupiter and Saturn formed first from the primeval ring. Resonances caused by their huge masses then caused the material in the ring further in towards the Sun to coagulate at the resonance locations. So the orbits of the outer planets, which formed first, may have caused the inner planets to form out of the ring at their resonance locations. It is at the moment a rather shaky hypothesis, but if the idea is true, we should be grateful to Jupiter and Saturn: for we owe our very existence to whichever of them supplied the resonance which caused Earth to form here.

Voyager 1 is now leaving the solar system. It is still in communication with the Jet Propulsion Laboratory and occasionally radios back information about conditions in the regions of space it is traversing. Its next encounter is in 41 000 years' time with a star in Ursa Minor called AC+793888. Meanwhile Voyager 2 is in transit to Uranus, which it will encounter on 24 January 1986, and Neptune, which it will encounter on 25 August 1989. By then, a first reconnaissance of the entire solar system will almost have been accomplished. Only Pluto will not have been visited by spacecraft from Earth.

Unlike the 1960s, the 1970s and early 1980s have been a period of severe budget cuts for NASA. The old justifications for space exploration, that the United States needs to win a space race with the Russians, or that space is a new frontier for America, have lost their capacity to win votes and money. Now, even a mission so obviously successful as Voyager is questioned. Voyager has cost about 250 million dollars. In return for that, the two spacecraft will, by the time the mission is over, have clearly shown us four planets and over twenty moons. Is it worth it: indeed, can the sort of discoveries made already by the Voyager mission even be costed? It is clear that the Jupiter and Saturn discoveries have no immediate economic benefit to mankind, although romantics might suppose that in the distant future we might harness the electricity of Io or collect the natural gas from Titan. Exploration is very difficult to justify on the basis of immediate economic advantage, and the long-term benefits are impossible to predict. Perhaps the only justifications are that the exploration of the solar system is one of the major events for which this century will be remembered, and that finding out about new and strange worlds is both fascinating and fun.

The one clear benefit that the encounters with Jupiter and Saturn have brought is a much clearer picture of solar systems in general. Jupiter has more mass than all the other planets put together. If only a little bit more mass had been added to it, it might have ignited. The pressure inside might have started a solar furnace, the nuclear fusion reaction which is at the heart of the Sun and which nuclear scientists are at present attempting to harness in fusion reactors on Earth. Saturn is smaller than Jupiter, but represents a likely model for the swirling origins of the solar system, with Saturn itself as the sun—star and the material orbiting around it in the process of differentiating into the various planets. So, in this sense, a planet, or at least a giant gas planet, is not so unlike a star. If Jupiter had ignited we might have found ourselves in one of those binary star systems that can be found so frequently in other parts of the sky.

Now, as we journey outward, we have no more information from spacecraft and have to rely on observational techniques from telescopes, orbiting satellites and other Earthbased equipment. In fact, for the long view, the wide-angle shot looking back at the solar system from outside, we would do well to look down a quite different instrument, the microscope.

5 The Message in the Rocks

Alec Nisbett

In the whole wide Universe there is just one planet that we can analyse in direct, physical detail. We can hold its substance in our hands, take it apart in the laboratory and tease out the history of its origins and the details of its present structure. Much of the evidence lies hidden very close at hand. Under a thin veneer of ocean and atmosphere, below a tenuous layer of life and its products, our own planet is mostly composed of rock. We also have a few fragments from the neighbourhood of Earth, some that have fallen as meteorites and some collected at enormous cost from our rocky satellite, the Moon. But the great bulk of the available evidence lies beneath our feet.

In geological time, rocks from Earth's molten interior form and reform, carrying different samples close enough to the surface for us to examine. Each individual rock carries the imprints of its experience, telegrams from the past, coded messages about the evolution of Earth. This is the story of their interpretation. It is also the story of a series of revolutions in human understanding. The first great conceptual advance was to understand that Earth does, in fact, have a history that is different from its present and that this can be read in the rocks that shape and underlie its landscapes.

John Sutton of Imperial College, London, has spent his working life studying the world's oldest rocks. For him, landscapes such as those of the Western Highlands of Scotland, have come to represent an enormous range of time-scales. He can look out over a valley that he knows to have been formed a mere eight thousand years ago, to an ocean, the Atlantic, which opened up eighty million years ago, leaving behind a narrow slice of North America brought here in a collision of continents that took place long before that. But all these events are recent history compared with the ages of the rocks beneath him, at nearly 3000 million years, some of the oldest in the British Isles. Traditionally, geologists have defined past ages by the fossils that lie buried in the rocks. Such classical methods have been remarkably successful in working out the history of what we now know to be the last 600 million years of geological time. However, outcrops of old Precambrian rock, features of the West Highland landscape, contain no fossils. When John Sutton began to work on it samples had to be sent to Canada for dating. Now, however, methods have been evolved in Britain to study that early period and the end result is a revolution comparable with that which took place in the nineteenth century, when geology really began.

That first revolution came with the realisation that it was possible to see back in time by looking down, through layer upon layer of sedimentary rocks. Some of the strata are composed of the remains of plants or animals – examples are the coal measures formed from trees and limestone from the shells of tiny sea creatures – others owe their origins to physical processes – the weathering of older rocks producing dust and mud which resolidifies. Later processes of erosion, such as the formation of the Grand Canyon by a river cutting into the strata, or man's own industry in digging mines sliced through the layers, so that by their order and their fossils they could be read. The deeper they lay, it seemed, the older they were.

Nineteenth-century geologists could see how long such processes took in modern times and from that they began to calculate the age of Earth. The rules they formulated worked marvellously well until they reached the original continental masses, rocks formed not in sedimentary layers but, as we now know, from beneath, like blobs of scum floating on the surface of a rocky sea.

At some places in the world these original continental rocks lie exposed at the surface. In such a landscape, depth and age are not directly related. John Sutton's Western Highlands provide one ancient example and another, older still, is in Western Greenland. In such regions some of the classical rules do still apply. For example, a volcanic flow that has intruded into cracks is younger than the surrounding rock, while fragments from an earlier period may have been enveloped and carried around by newer materials. These concepts are simple and logical enough, but it also seemed at one time reasonable to suppose that the rocks of the deep ocean floors must be older than the continental rocks that were raised above them. A turning point in modern geology came with the demonstration that the reverse was true – sea floors were mostly fairly young, compared to the majority of continental rocks.

Cooling Rock

Continental rocks, like those of the ocean floors, are mainly composed of mixtures of mineral salts that have solidified to form interlocked arrays of different kinds of crystals. The nature of the rock is determined in part by the chemical composition of the original molten mass. As the temperature falls, some components with a strong chemical affinity and high melting point will be the first to crystallise, then others in turn until the whole rock is solid. The size of the crystals will depend on the rate of cooling: rapid cooling, usually near the surface, produces small crystals with a fine structure; slow cooling leads to larger crystal domains. If a rock sample is cut across and polished and then rotated in polarised light under a microscope, the crystals appear afire with an everchanging pattern of brilliant colours. Already we have one set of messages carried by the rocks – their chemistry and thermal history. They may also show evidence of reheating and deformation. But in a way, all this is obvious: their deeper message was far more subtle.

Long after they have solidified, the rocks of Earth have continued to pass heat to the surface. It filters upward very slowly, so that an area the size of a football pitch transmits barely enough energy to illuminate a light bulb. But, in geological time, that still adds up to a tremendous amount of heat. Where did it all come from? How long will it take for Earth to cool? And what does that tell us about the age of Earth? In Victorian times this last question led to one of the biggest disputes in scientific history.

It was between, on the one hand, strong, observational geological evidence for the great antiquity of Earth and, on the other, equally solidly based physical analysis which raised an apparently insurmountable obstacle to that antiquity. And the leading proponent of a 'young' Earth was Lord Kelvin, one of the most brilliant scientists of his time. Kelvin worked out the thermal history of our planet, assuming that it started out as a molten body, and from his measurements of heat flow was able to calculate that a body of this size would solidify and reach its present state in not more than twenty or thirty million years.

His opponents were highly skilled, practical geologists, men who believed they understood how sandstones and limestones and other sedimentary rocks had formed. They could see just how slow were the processes which made such rocks today and so could estimate the time required to form the enormous piles of sediments they had seen in many parts of the world. There was no way that they could fit such an enormously protracted series of events into the tiny compass of time that Kelvin offered them.

Neither side could see any flaw in its own argument and neither could properly comprehend the arguments of the other. It was a total impasse: neither could win and neither would give up. What none of the protagonists knew, was that Earth has its own internal source of heat. The discovery of radioactivity and its implications for geology were enormous. The first and most direct result was that Kelvin's conclusion fell: Earth really could be ancient. But how ancient? Again it was radioactivity that provided the answer: a direct and independent measure of the age of rocks. This is the most important message a rock can carry – its date of birth.

The Rocks tell their Ages

In the mineral stew from which most rocks are formed there are usually a few radioactive elements. For some still to be around after tens to thousands of millions of years, such elements must decay very slowly; accordingly the useful radioactivity will be very faint. But it is not the direct measurement of decay that is required: simply a comparison between the amounts of the parent elements and daughter elements that result from the decay.

A given radioactive element will decay at its own characteristic rate. After a given period of time, the half-life, half will have turned into a daughter element; after a further similar period, half the remainder will have gone, and so on. The half-lives of suitable pairs of materials can be measured in larger samples in the laboratory. So what is then needed is a count of the two sets of atoms in a suitable sample of rock.

In the Age Laboratory in the Department of Geology of Oxford, Steven Moorbath used a pair of elements, an isotope of the relatively rare rubidium, which decays, with a half-life longer than the oldest rocks on Earth, into strontium. The technique used is to separate the materials chemically, then to sort them by mass. On a glowing filament, the strontium is heated and is emitted as an ion, an electrically charged particle which is drawn away by the electrical field of a mass spectroscope. The moving particles then pass through a strong magnetic field, so that the lighter the particle, the more its path is bent. A sufficiently powerful field will separate particular isotopes from their neighbours; only one at a time will be detected, and its relative frequency measured. From this, in due course, its ratio to the parent rubidium is also calculated, and from that the age of the rock itself.

Some of the oldest known rocks on Earth appear at its surface in West Greenland. Radioactive dating of the banded rocks on the right gives an age of 3700 million years. The broader white strip is an intrusion of molten magma from 2500 million years ago.

Steve Moorbath finds that some of the rocks he has collected in West Greenland have an age of well over 3700 million years, making them some of the oldest, or earliest, continental crust known anywhere on Earth. And yet, to all outward appearances, these rocks are almost identical to others that he has collected in Chile, rocks so new that they must still be forming today. So the process of continental rock formation has taken place over a period of at least 3700 million years.

Is there anything older? One candidate in Moorbath's collection comes from a banded iron formation in a mountain at a place called Isua in that same region of Greenland. Only a few small fragments of volcanic rock could be found trapped within the sediments, but in one of them were eight tiny crystal grains that between them did contain enough of the radioactive decay product to measure. It dated at 3824 million years, plus or minus 10 million.

From this we can see that there were both volcanic and sedimentary rocks in those early times – and also a process of erosion to form the sediments. This means that there was not only cloud but also liquid water at the surface, and that water evaporated and rained much as it does today. Then, within a few hundred million years of the formation of those oldest rocks – and in geological terms remarkably early in Earth's history – life had appeared. It was feeding on the carbon dioxide in the atmosphere and pouring out what was then a dangerous poison, free oxygen, in our World's first major episode of air pollution.

The Gap in the Record

We now have an early picture of Earth, one that is not so different from today's. But it takes us back less than 4000 million years. Is there a hidden period that lies beyond that, an early Earth of which little evidence remains? What happened between the formation of Earth, whenever that was, and the production of its present, oldest rocks? One further line of enquiry makes use of other kinds of radioactivity, much of it in the deep-formed rocks of Earth, but also some in the rocks that fall to Earth from space, the meteorites.

Uranium is the key. One isotope of uranium decays very slowly indeed, then passes relatively speedily through a series of intermediate stages, to finish up as a particular isotope of lead. This provides a second means of dating not only the oldest Greenland rocks, but also meteorites, and it is found that there is a difference between the two. Some of the rocks from space are over 700 million years older than any known on Earth. But measurements of the relative abundance of lead isotopes indicate that the materials in the meteorites and of the Earth itself have some kind of common origin nearly 4600 million years ago. It is now believed that Earth was formed soon after that, perhaps 4500 million years ago.

It has been proposed that the solar system began as a rotating cloud of hot gas: the Solar Nebula. In that, particles of dust condensed and began to stick together; then as the grains grew in size they sank to the central plane of the spinning nebula. In this much denser sheet, gravitation was able to pull the material together into larger chunks, perhaps a kilometre across . . . or so this leading theory has it. And everything so far could happen within a few tens of thousands of years.

These lumps would then continue to grow by gravity alone. This would take much longer: for Earth, about 100 million years. It would be an epoch of intense bombardment, and each missile that struck Earth would add to its heat as well as its mass: gravitational energy would become kinetic energy; this in turn would be converted into heat enough to melt both rock and the iron that is found in many meteorites, until the whole body reached a temperature sufficient for the iron to segregate and sink to the centre of Earth, releasing yet more energy and heat. Even without any radioactivity at all, at this stage in its history Earth would be very hot. It would be covered by volcanoes and by an atmosphere formed by the carbon dioxide that poured from them. However this is only a theory, albeit widely held.

The Poor Man's Space Probe

To study the evidence about events that occurred in the neighbourhood of Earth 4600 million years ago, we do not have to search very far. Earth itself is the probe, sweeping up or simply getting in the way of rocky debris that lies scattered in space. If we simply wait, it will come to us. Some of these ancient bodies are still arriving. The famous meteor that has left its mark near Flagstaff, Arizona, hit the ground at about 17 kilometres per second, releasing enough energy to vaporise its 3000 tonnes of iron on impact. But smaller fragments decelerate in the atmosphere and land with the cold of space still on them. Surprisingly, small meteorites are cool enough to pick up straight away.

Ed Olsen, of the Field Museum of Natural History in Chicago, has a collection of meteorites, some of which were simply picked up by laymen and others that he has found himself. A good place to search for them is Antarctica, where meteorites lie on or are carried in the ice and reappear in certain favourable places. A Japanese team, having discovered this, collected many hundreds.

Meteorites come in a variety of forms. Many are simply stone-like materials rich in silicon, or chunks of iron. On the face of it this suggests that they may be fragments from a planetary object which broke up at some time in the past and have been whirled around the solar system until they eventually landed here. Their nature is consistent with the theory of segregation by heat into a heavy metallic core with a lighter rocky mantle outside it. Their source therefore seems easy enough to explain, except for one problem: if the parent bodies for the rocky or metallic (or, in fact, sometimes mixed) fragments were much smaller than Earth they would need an additional source of heat to melt them; gravitation alone would not provide enough. There seemed to be some evidence that this was in fact the case, that many of these objects had not come from one or a few large bodies of the size of our Earth or Moon that had then broken up, but instead from objects of much more modest size, which had still, somehow, managed to segregate an inner core from their outer mantles.

The message in these rocks from space is enigmatic. Does it support or deny our hypothesis for the formation of the early solar system and Earth's place within it?

The Message from Allende

The rocky, metallic and mixed meteorites are older than any rocks on Earth, as calculated by the usual radiometric dating techniques, but not so old as a further, apparently unrelated type of meteorite of which the most celebrated example fell one night in 1969 near a small town in Chihuahua, Mexico, called Pueblito de Allende. Over the years, fragments of the Allende meteorite totalling several hundred kilograms have been recovered. Allende is what is called a 'carbonaceous chondrite', which means that it is composed mainly of a dark grey claylike substance that picks up carbon and organic compounds from the dust and debris of space. Carbonaceous chondrites do not appear to have gone through the processes of planetary segregation into iron and rock; they are altogether more primitive. Some have inclusions, pinhead grains of a very hard, white mineral, and until Allende nobody knew what these grains were. There had never been enough material to study.

In this meteorite, the grains were not just pinheads, some were the size of small pebbles. The techniques for their analysis were, as luck would have it, just being developed and for once, as one investigator put it, 'you didn't have to kill a museum curator to get some'. Allende's curious white lumps attracted the special interest of Larry Grossman of the University of Chicago, for it turned out that they had a very high melting point and contained just such minerals as he had predicted would condense first from the rich, hot gases from which the solar system must have been made. So it seemed reasonable to his colleague Bob Clayton to make a routine check on that by using a technique which would indicate the temperature of the gases at the time when they had condensed.

Again a mass spectroscope was used, though this time it could be smaller and simpler since it was measuring the relative abundance of the isotopes of a light element, oxygen. Not only are light elements deflected more easily in a magnetic field, but in addition their mass ratios are greater, in steps of about 6 per cent oxygen. The plan of the experiment was to discover the ratio of the isotopes oxygen 16 and oxygen 18, which condense at different rates in a manner that is dependent on temperature. The normal ratio of these isotopes is well known and it could be assumed that there would be no difference in this case. They could guess the temperature and so the proportions they would find. But, as it turned out, the ratio they actually did find was totally different, in fact so different that any effect of temperature would be completely swamped. The oxygen was peculiar in a way that at first they could not understand. It was as though it had been produced by some different process and perhaps in some other place from any kind of oxygen previously encountered.

The apparently crazy result was noted by another group, a team at CalTech in California, which had orginally been built up to study moon rocks in ultra-clean conditions. (Hence the affectionate title of their laboratory – 'The Lunatic Asylum' – which did not amuse some senior CalTech administrators when a stream of important papers was credited to this small enclave within their august institution.) In the Lunatic Asylum they, too, had a few pieces of Allende and were using it to look for the missing source of heat, something which would melt and so segregate the materials of aggregates which had grown to less than planetary size. What they wanted was a radioactive source of heat.

Among the sources that theoreticians considered possible was a rare radioactive isotope called Aluminium 26. This decays into Magnesium 26, giving off energy, heat, with a half-life of 700 000 years. In Allende they looked for crystals that might reasonably have started life as compounds of aluminium but which would also contain an unusually large amount of Magnesium 26. And with their mass spectrometer (cheerfully labelled 'Lunatic 1') that was exactly what they did find. The additional heat source was triumphantly identified by Thyphoon Lee, Dimitri Papanastassiou and team leader, Gerry Wasserburg. With some glee, Wasserburg commented:

> The concentration of this element was so enormous that it would have been the dominent source of ionisation of gases in the whole Solar Nebula at the time the solar system formed, and secondly, a sufficient heat source to cause the melting of a planet only two kilometres in radius. The intensity of the heat source is not just large, it's overwhelming. For two decades people wanted to know what melted planets, and now we've proclaimed that we think we know what it was, it's an embarrassment of excess. Now people say, haven't you got too much?

The heat source, the oxygen anomaly and other discoveries that followed all point to a most remarkable sequence of events. Somewhere near the gas cloud that was ready to form our solar system, an old, dying star blew itself apart, rather as the Crab Nebula was seen to, by the Chinese, in AD 1054.

This local supernova would have happened about 4600 million years ago, seeding our own region of space with the elements from which our planet, its rocks, its metallic core and all its living forms are made – and also with the aluminium heat source. To avoid making more of a coincidence of all this than is necessary, it is reasonable to suppose that the supernova also triggered the collapse of the material that it seeded. The crucial formation processes would have been complete well within the million years or so of abundant radioactive heat.

So now we have a plausible theory of the formation of the solar system, strongly supported by evidence gathered from meteorites, and we also have a partial history of our own Earth's rocks. But we still have a gap between the two of hundreds of millions of years. Fortunately there is one other source available to us which can help to fill that in.

Moon Rocks

From the cratered moon, the US Apollo and Soviet Luna programmes brought back the rocks of different regions. Again they can be dated (cautiously, given the limited availability of the materials obtained at such fantastic cost) by radiometric techniques. Grenville Turner at Sheffield University uses the slow conversion of a radioactive isotope of potassium into the gas argon. This measurement is trickier and more difficult to interpret than some of the others that have been used, because the gas argon remains trapped in the rock only so long as the rock itself remains impermeable (any impact that heats up the rocks will release it). In consequence, it gives, not the full age of the solid rock, but rather, that of its most recent violent event. Using this method for meteorites, there is again a tight clustering of ages around 4500 million years, with the oldest at 4550 million. The Moon also yielded samples of native rock brought back by Apollo 17, which three laboratories, including Wasserburg's, dated to 4550 million years: the Moon was evidently formed at much the same time as the rest of the solar system. But most of the rocks from the Moon's surface show ages clustering around 4000 million years, which makes even those substantially older than any found on Earth. Their internal clocks were started – or set to zero and restarted – by some process or event about 500 million years after the formation of the solar system. Wasserburg's group called this 'the Terminal Lunar Cataclysm'.

'The conclusions,' Wasserburg says, 'are perfectly clear. Earth and Moon were bombarded by fairly sizeable objects, fifty to a hundred kilometres across and going really quite rapidly, which smashed up the surface of the Moon and must certainly have smashed up the surface of Earth at this late time.'

So, it seems, Earth's own record of events from that time would have been erased, by its own crustal activities, while on the Moon, the cooler, more solid crust retained it. Apart from that, studies of the Moon serve only to emphasise not how similar, but rather how different our histories have been. In particular, Earth, quite unlike the Moon, has developed a vast iron core. But that paradox, too, is playing its part in our growing understanding of the evolution of Earth to its present form.

Earth's Core

Much of what we knew about the interior of Earth is collected from the networks of seismometers that have been established to record and measure earthquakes, and to monitor nuclear tests. A seismometer usually consists of a system of masses that are mounted loosely enough to stay where they are if the ground moves beneath them. They will register any vibrations, including passing traffic or a technician's footsteps which, fortunately, are easily recognised and can readily be discounted.

Any big earthquake sends waves right through Earth as well as around, through the surface layers. Earth's core provides a discontinuity in density, so that as it passes through, the wave changes direction slightly. By analysing the pattern emerging at the far side, the size of the core can be calculated.

Determination of its shape is more difficult. Certainly, there appear to be blobs of a denser mantle material above the core—mantle interface, and that boundary itself may not be smooth, according to fluid dynamicist Raymond Hide of the British Meteorological Office at Bracknell. He speculates that bumps on the interface about a kilometre in height could account for otherwise puzzling phenomena to do with Earth's magnetic field and its occasional, irregularly timed polarity reversals – the occasions when Earth's magnetic north and south poles change places. There have been several dozen reversals in the last two hundred million years; while in the Permian era Earth was locked in a single polarity for some fifty million years.

In addition there are large-scale gravitational and magnetic anomalies, detectable from Earth's surface, which might be related in some way to bumps on the core.

How such irregularities might be formed is unclear. Differential pressures from above might play a role; so could radioactive heating within the core itself. All that can be said for sure at present is that the core remains a domain of mystery. We do not even know how its supposed dynamo generates the magnetic field with which we are so familiar, though it is reasonable to suppose that fluid-flows in or around the core play an important part. What we can see, however, is the effect of the magnetism at the surface. Indeed, that provided one of the first and most important clues to the evolution of the Earth and, by implication, the other rocky moons and planets in the solar system.

Fossil Magnetism

When a rock solidifies, any traces of magnetic material within it retain the imprint of Earth's magnetic field at that time. A series of rocks that solidified at different times will carry the record of the ancient magnetic fields. A strip of rock which has gradually solidified from one end to the other acts like a vast tape recorder magnetising the strip first one way and then the other as the poles change places. The sea floor is such a strip. This provided one of the early pieces of evidence which led to the now familiar picture of ever-shifting landmasses, the slow waltz of continents as a kind of superficial scum on the surface of moving convection cells in the mantle beneath – the system known as plate tectonics.

Less familiar, however, and much more recently investigated are questions about how the whole system is driven, and how the old continental rocks with which our story began came to be separated from the young rocks of the sea floor and from those of the mantle; in fact, how the scene was set for the fantastic varieties of mineral forms that we find scattered around Earth, in its varied range of landscapes.

Creating Continental Crust

Earth's mantle contains its own store of heat-producing radioactive materials, mainly isotopes of uranium, thorium and potassium. Because rock conducts heat away rather poorly, it

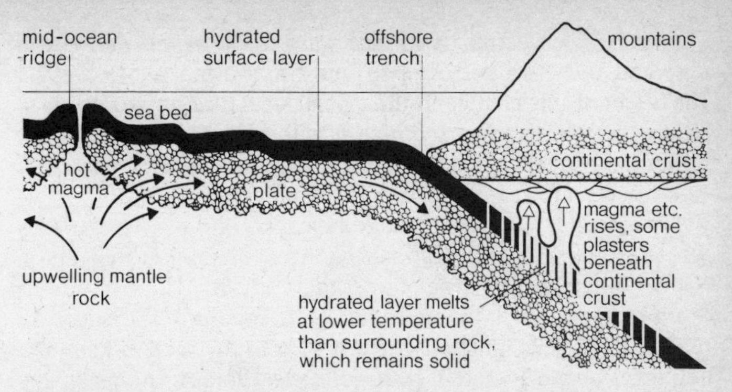

Sea floor spreading and the formation of continental rock are linked. The key is water: hydrated rocks melt more readily than the material around them. In addition, Volcanoes appear both at the mid ocean ridge and in the continental margin.

gets hotter, expands, and the less dense masses tend to rise, leading to a slow viscous overturn of what behaves like a thick, sticky fluid. If it is difficult to imagine rock, a brittle solid, flowing like this, consider any glacier. There, the sluggard flow of another brittle solid, ice, is obvious.

A widely accepted idea of how continents grow runs something like this: the hot mantle wells up at mid-ocean ridges, cools and spreads away from the middle until this heavier basaltic rock meets the lighter granitic continental material and is deflected beneath it. The cooler plate thrusts downward at an angle through the mantle. Earthquakes monitoring sensors can detect fracturing along the planes of contact between the descending slab and the mantle. According to the standard story, this causes friction enough to heat up and melt the descending rocks in those regions, to form pools of molten magma. This threads or forces its way up to the surface as volcanoes, adding new rocks to the continents. A problem with this is that rather a lot of heat would have to be generated in just the right places.

John Tarney of Birmingham University has studied rocks from sea floor dredging and drilling expeditions and has also compared granitic continental crust of different ages, including some formed in the last 200 million years along the western side of the Andes where continent formation continues today. He finds that their chemistry suggests a somewhat more elaborate process. He believes that the key is hydration of the mantle rock as it comes into contact with sea water, first at the mid-ocean ridge and then as fissures open into the moving sea bed plate. One effect of this change in the crystal structure of the rock is to reduce the temperature at which it will melt. So, as the plate is subducted beneath the continental crust, the hydrated parts of it melt much more readily than the surrounding material.

When, in due course, it resolidifies, a second process governs the order and eventual destination of its component molecules. In any molten rock, Tarney finds, there is a tendency for atoms with a larger atomic volume to be squeezed out of the matrix of solidifying crystals, pushing them back into the melt. There are, therefore, proportionately more of these 'larger' atoms in the magma that wells up through the surface as volcanic lava. (The volume occupied by an atom is not directly related to its atomic weight, so it is not simply the heavier atoms that are segregated in this way.) These 'incompatible' elements include potassium, rubidium, uranium and thorium, carrying with them a disproportionate amount of the radioactivity from the mantle. This supplies the extra heat.

The more fusible minerals, those which have a low melting point anyway, include oxides of sodium, aluminium, calcium, iron and, from the hydrated rocks, silicon. These and some of the incompatibles seep upwards. Since some of the magma is recrystallised before it reaches the surface, the melt is subject to a third stage of enrichment on the way. Some material plasters itself to the underside of the continental masses, the rest either lines the volcanic vents or erupts onto the surface. The mantle that remains in circulation below is left with ever more of the minerals that have higher melting points; it becomes increasingly rich in magnesium and poor in silicon.

This process has been going on for a long time, so it is reasonable to ask whether some of the material now coming up from the mantle has been round its full cycle once or several times before. Tarney, comparing the South American granites with those of older continental rocks, finds just such a progression. That which emerges later is, indeed, already partly depleted in the incompatibles. Much of his evidence for this is drawn from precise measurements on rare earths, a series of elements that have different masses but identical chemistry. Their most important characteristic for his purposes is that their atomic size, and therefore compatibility within a crystalline structure, is markedly different over their range. He finds that over the history of Earth's mantle there have, indeed, been changes in the proportions of rare earths emerging from the depths. Those that are most incompatible have come out in greater relative abundance in the earlier continental rocks, and less in the later.

A Message from the Mantle

It would be interesting to check some of this by digging a hole through the sea floor or continental rocks to find out whether the mantle really does have the expected composition. The easiest point for such an exploration would be where the surface plate is thinnest – below the sea. Such an experiment is technically feasible; it has been proposed and planned, but then abandoned on the grounds of cost. To drill the continent would be even more prohibitive, because seismic soundings suggest that these plates, together with whatever may be plastered beneath, are very much thicker. Fortunately, however, such an experiment has already been performed for us by Nature herself.

Pockets of gases trapped beneath the continental crust have in some places forced their way through to the surface and in their rush have carried up materials that are quite different from those found elsewhere. All the world's natural diamonds come from such kimberlite pipes – so called after a famous example in South Africa. Some, in other parts of the world, offer no diamonds, but instead a harvest of data that is rich beyond price. Each kimberlite pipe is a direct line, not just to the mantle but to a particular moment in its history. And these messages from the mantle are brought up to the surface and simply left lying around for the geologist to pick up. Indeed, in one place in the United States, it is first picked over by ants, which gather up heaps of insect-sized boulders to protect their nest from being washed away by rain.

Keith O'Nions, a British geochemist working at the Lamont-Doherty Geological Observatory in New York State, has sifted through such debris for particular tiny telltale fragments that contain long-lived radioactive isotopes of neodymium, lead or strontium, together with their decay products. Each sheet of his data is a page from Earth's chemical history which can be compared with mantle rocks that emerge today at a mid-ocean ridge. It is found that in the latter there is measurable reduction in rarer elements that must already have been segregated into the continental crust during earlier cycles within the mantle. It's an irreversible process: it all goes one way. So it seems possible that Earth's surface began as mostly plain sea floor, with a few blobs of scum being the proto-continents, and that it will be increasingly covered by the more complicated and varied continental crust. There is a sense in which the geological evolution of Earth parallels the evolution of life itself. At each stage, geophysical and geochemical processes favour segregation into the wide range of separate forms that we see.

One problem remains. The process here described sounds as though, in principle, it should be continuous. In fact, it turns out that the continents have grown in fits and starts, with periods of rapid accretion interspersed with times of no growth at all. There have been about five major episodes; starting with that which began some 3700 million years ago, and the latest of which continues today. Can this intermittent growth be accounted for by the staggered, stumbling waltz of continental drift, with its own hesitations, collisions and shifts? That is still debated. John Sutton sums it up:

> Sooner or later, as the facts accumulate, there's still the opportunity for someone to come forward with an explanation for such things as the bumps on the core and the spurts of continental growth, and to bring these together in a hypothesis that will actually explain the mechanism of the changes which have occurred during the evolution of Earth. But we have already seen one revolution, and have caught a glimpse of revolutions to come.

Earth is one planet among many: we know in some detail the elements from which it was formed and how it evolved to its present state. We can also put a date on it; it is 4500 million years old, approximately. We can fit all this data in with the less detailed data we have from the other planets. The picture that emerges is one of the solar system forming out of a single cloud of gas and dust at about the same time. Exactly when the Sun ignited is not known, but all the planets, and the Sun, formed more or less together, with the denser, inner planets perhaps forming first.

But where did that initial cloud come from? It was a pretty sophisticated cloud. It was not composed of the simple elements that one might expect to find floating around in space. Because of what we have discovered from Earth and the meteorites that have hit it, we know that there were elements composed of atoms with heavy nucleii, such as iron, gold and uranium, already in that cloud. These heavy elements could not possibly have been formed except inside a very hot furnace – far hotter than the centre of Earth or any man-made furnace on the surface. Something like the interior of a very large star, perhaps, which has since exploded, releasing those elements, would fit the hypothesis.

With this in mind, we can now journey outwards from the edge of our own solar system, towards an object which was to be one of the most remarkable 'finds' in the history of astronomy.

6 The Crab Nebula

Alec Nisbett

To astronomers, the Crab Nebula is special: they talk of the astronomy of the Crab as though it were different from the astronomy of everything else. Rules learned from the observations of other objects do not seem to apply here. As each new technique is devised and turned on the Crab, it reveals something unexpected until, eventually, surprise itself becomes routine. The story of the Crab has been said to be 'more like a tale of mystery and imagination than a sober, scientific recital.'

Alternatively, we may think of it as a classic detective story, full of twists and turns in the plot. It tells of an investigation into the cataclysmic death of a star; a post-mortem in which the body lies some 30 000 million million miles distant, remote enough for any Earthbound pathologist, but in astronomical terms a modest span, 6000 light years, only one fifteenth of the diameter of our galaxy. Despite the distance, its examiners have reported with growing amazement that it provides a vital link in another mystery story, that of our own existence. It may be that we live to observe the Crab only because such a stellar death seeds its neighbourhood with the raw materials required for the creation of biological life and of the physical world that we inhabit. According to Philip Morrison of MIT:

> The Crab Nebula, somehow, by good luck, is close enough to us and is a complex and new enough event to act as a kind of Rosetta Stone to the life and death of stars. The trouble is that like the Rosetta Stone we don't know very well any of the languages in which it is written – but we try to translate them together, and we are making some progress.

Nearly all of the evidence available to us has come from the study of the original star's scattered and decomposing remains but, as in any inquiry into a violent event, our first question should be, did anyone see it happen?

A Guest at the Court of the Emperor

In our story the first clue comes from libraries of oriental books, and is found in ancient Chinese chronicles. By our calendar, the year was AD 1054 when this entry was made:

> The Chief of the Astronomical Bureau reported that from the fifth moon of the first year of the period Chih-ho a guest star had appeared in daylight in the Eastern heavens.

In the Sung Dynasty such an omen was a matter to be explained to the Emperor. The formal words of Yang Wei Te are quoted in a second chronicle:

> Prostrating myself: I have observed the appearance of a guest star. The star was of a somewhat iridescent yellow colour. Respectfully, according to the dispositions for Emperors, I have prognosticated and the result said, 'The guest star does not infringe upon the star Aldebaran; this shows that a Plentiful One is Lord and that the country has a Great Worth.'

More helpfully, the chronicles report the region of the sky in which the guest star shone, the constellation of Taurus, and give another subjective account of its appearance: it was said that pointed rays shot out from it on all sides and that the colour was reddish-white. And a precise number is given, one that is of direct use to modern astronomers: the guest star was visible in daylight for twenty-three days.

In 1895, when the Yorkshire carpet manufacturer Albert Crossley donated a 36 inch reflecting telescope to Lick Observatory, to be installed on Mount Hamilton in Northern California, the sky outside the solar system was widely regarded as unchanging in historical time. In 1899, long before that idea was finally abandoned, the Crossley Telescope was turned to the Crab. After many hours' exposure, a photograph was taken that, even by today's standards, was excellent. It shows an irregular cloud containing dense and lighter patches. If the same telescope and techniques are used today, the new photograph will show a similar structure, but spread over a larger volume of space, as marked by the unchanging pattern of the stars that lie in the same field. In a human lifetime, the Crab has visibly grown.

When these changes in its appearance first became evident in

Two stages in an explosion. The Crab in 1899 (above) *and today. Both views were obtained by making a long exposure on a telescope of modest size, the 36-inch Crossley telescope at the Lick Observatory in Northern California.*

the photographs, astronomers were reluctant to ascribe them to physical motion, but in the end that was the only possible interpretation. The cloud is expanding at a rate of nine hundred miles every second, or three million miles an hour; successive images form a time-lapse sequence showing the latter stages of an explosion that might have happened in the eleventh century or thereabouts. In fact, if the present motion is extrapolated backward, it seems to have started in AD 1150, not 1054, a slight difficulty. But it is in the same region of the sky as the oriental guest star, so astronomers have made the reasonable assumption that the Chinese did see the same explosion in its brilliant, early stages.

Sadly, other records are sparse and insecure or simply non-existent. For example, some American Indian cave paintings might just possibly show a bright new star shining near, or with brightness comparable to the Moon. But the Arabs, who were good astronomers, and the Normans, who depicted the 1065 appearance of Halley's Comet in the Bayeux Tapestry, have left us no account that has yet been found of any precursor to the Crab.

Modern astronomy records many events in which a star flares up temporarily as a shell of gas is thrown off – these are called novae – but the Crab is in a different class. To have been visible by day for so long and at such a distance, the explosion was a great deal larger, in fact, as it now appears, the almost total disintegration of a star much larger than the Sun. They call it a supernova.

The Nebula Discovered

In the eighteenth century many possessors of telescopes of modest size lent an eye and a hand to mapping the skies. In the 1740s an English amateur, John Bevis, produced maps of high quality, embellished with yellow, four-pointed stars and pictorial representations of the constellations. In one horn of Taurus the Bull, he placed a grey disc, his symbol for one of the faint, nebulous patches of which a few hundred could be seen with a small telescope.

Then on 28 August 1758, a Frenchman, Charles Messier, observing from a tower above the Hôtel de Cluny in Paris, discovered it afresh. Ten years later, he placed it first in his catalogue of just over a hundred special objects. It became 'M1'. But this pride of place is illusory: Messier had no interest at all in

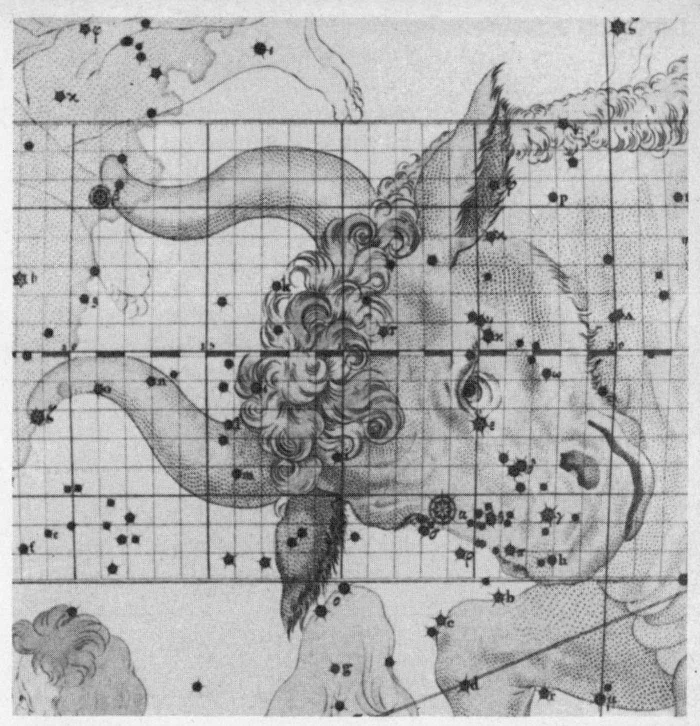

The earliest map showing the position of the Crab was drawn by John Bevis in the 1740s. It appears as a round rosette-shape – different from the four-pointed stars – at the tip of the Bull's lower horn.

static stars or nebulae when there was so much movement in the sky. The return of Halley's Comet, as had been predicted, and for which Messier had been searching when he rediscovered the Crab, had caught the public imagination and galvanised science. To make his name, an astronomer of that period had to find a comet or, better still, a new planet or asteroid. A comet is easy enough to spot when it is near the Sun and has a tail but, as it first appears, it is just a faint grey patch. The first thing, therefore, which an observer needs to know is whether it really is something new, or whether it was there the week before. Messier's generous service to his fellow comet hunters was to publish his list of all the main objects that could confuse them in their search, fuzzy patches of which he could tell them, 'Don't look here, it's not a comet. It won't move!' Our cloud of stellar debris thus became number one on a list of objects to be ignored.

In Ireland in 1844, William Parsons, the third Earl of Rosse, used his 36 inch reflector to sketch what looks remarkably like a lobster or a pineapple, and then gave it its present name, the Crab. It does not appear very much like the object of the photographs made a few decades later, either; it is not one of Rosse's better efforts.

The Structure of the Crab

Even with the finest telescope, the Crab does not reveal much more to any other naked eye than it did to Rosse's; there is a little more structure perhaps, but without sufficient brightness for the eye to register colour. For that, a time exposure is required. One photograph from a 200 inch reflector, taken in 1950, shows an amorphous pale blue-green cloud wrapped in brownish-red filaments. How true is this representation? Ordinary colour film is imperfect in its rendering of different hues, but this does not usually bother us because our eyes are used to great variations in both natural and artificial lighting. In everyday earthly colour photography the most noticeable error is in the representation of flesh tones, but these can easily be corrected by eye. Here on Earth, we know how to allow for such things, but we have no direct experience of objects in space with which to compare the photograph. In any case, the Crab is unique.

In the late 1960s a different technique was used to create a more accurate colour rendering based on the additive process devised by the Scot, James Clerk Maxwell in 1861. On the Crossley Telescope three separate six-hour exposures were made in black and white, but using different broad-band colour filters. The three images were then added together to give a new composite. The earlier photograph is more dramatic; the later, truer balance is more delicate and beautiful. In particular, the amorphous region appears to have a much softer electric blue, shot through with the reddish-brown filaments. Stars within the image are seen more clearly, but most may be presumed to be unrelated, located by chance along the same line of sight. However, near the centre of the gas cloud there is a region of lower intensity and there are two stars 'within' that. It had long been suggested that one of those might be associated with the Crab, but there seemed no way to confirm that, and in any case the main candidate did not seem bright enough to account for the glow in the gas around it.

Such a photograph shows the way to the examination of particular features, for example, the filaments, in greater detail on larger telescopes, such as the 120 inch at Lick. For Joe Miller of the University of California at Santa Cruz:

> There is one very important reason why we are studying the filaments in the Crab Nebula. They represent material that was just recently inside a star; material that has been greatly enriched by the heavier elements that is now being returned to the area between the stars, and at some time in the future may make itself into new stars. This process goes on again and again in the Universe, with old stars giving the material to provide for the birth of new stars at some time in the future. We ourselves and the world we see around us were at some time in the past perhaps involved in one of these supernova explosions.

If it were, then in the Crab we see a crucial stage in that process. In the original star, it is supposed, hydrogen nuclei combined to form helium and the star shone brightly radiating the energy of nuclear fusion. When the hydrogen ran out, other fusion reactions took over, turning helium into carbon, oxygen and finally iron, all of which we need for the atoms and molecules of our own flesh and blood. However, according to the findings of nuclear physicists, iron lies at the bottom of an energy well; when lighter atoms combine there is energy to spare – but the iron nucleus is at the end of that line. Beyond iron, energy must be added to fuse larger nucleii and they give off energy again when they split. Many of the heaviest elements, including all above the size of the lead nucleus, are unstable: they break up spontaneously, releasing the energy that was required for their formation.

Fusion of the lighter atoms could account for most that we see in ordinary stars, at some stages releasing energy from the hottest central regions so fast that the radiation pressure would inflate the cooler outer parts like a giant balloon. The star would form and reform in different sizes and colours during each stage of fusion and, in the case of the Crab, at some point in this process it exploded. To try to find out what stage had been reached, the composition of the debris was examined. It appeared that they did, indeed, contain a cocktail of elements, including the raw materials needed for life and for the worlds on which it could evolve. The former star appeared to have run through a wide range of nuclear element-building processes before exploding. To the optical astronomer, then, the Crab has a power to explain the evolution of stars and of the galaxy as a whole that is out of all proportion to its appearance as a cooling ember of its former fires.

The Radio Beacon

The optical glow is far too faint to see with the naked eye, but with the birth, now over thirty years ago, of the infant science of radio astronomy the Crab shouted its importance from space. This was totally unexpected: though it was so distant, the new telescope saw it as one of the brightest sources of radio waves in the sky. Initially designated 'Taurus A', the radio source was identified with the Crab by John Bolton who recalls his early days in this field:

> For my first two years in radio astronomy, the conventional optical astronomer did not, in fact, believe that the radio signals we were receiving were from outside Earth at all. I had spent my time during the war as a radar officer in the Royal Navy. I spent parts of it on research and parts of it operationally; and one of the things which changed the optimum range at which we could detect a German aircraft was whether the Milky Way was above the horizon and in the aerial beam or not. So to me, extraterrestrial radio emission was a very genuine thing. I looked at this radiation from outer space with just the same belief that a conventional astronomer looks up and sees stars and believes they are there. After the war I got the opportunity to try and split up this broad band of radiation into discrete objects.

With a group of what looked much like today's UHF television aerials, Bolton located three radio sources in different places in the sky. Two had no obvious optical counterparts and attempts to identify them with faint galaxies met with scepticism, but the third was at the same place as the Crab, an object that was widely regarded as odd enough for this new observation to be more readily accepted. By placing his receiver on a cliff above the sea, Bolton was able to compare a direct signal with that coming by a second path, reflected from the sea. This enlarged the effective aperture of his instrument to twice the height of the cliff, giving him much greater direction and precision than would otherwise have been possible with such a simple instrument. The Crab helped to establish the value of radio as a source of new ideas and so to pave the way for the huge dishes and more complex interferometers that were required to resolve the fainter details of the Crab's radio picture, and then to map their structure.

An Accelerator in the Gas Cloud

Bolton's unaccountably powerful radio signal can now be mapped in detail and is seen to be coming from regions that coincide with the amorphous clouds. However, in radio it emitted not just a glow, but a blaze – pouring out radio waves with an intensity a hundred times greater than our own Sun. For a cloud of gas that had been cooling for 900 years it was far too bright; something must be keeping it hot, still supplying energy to it. The cloud was being acted upon by some completely unknown influence.

There was further evidence of this in the visible light from the amorphous region of the Crab. When seen through polarised glass of the kind used in some sunglasses, the image changed as the filter was rotated. An ordinary glowing mass of hot gas would not do this; it would be the same however the glass was turned. So astronomers looked for a mechanism that would produce both light and radio waves and that could also polarise the light, organising it in this particular way.

Just such an effect had been seen on Earth and was suggested for the Crab by a Russian, Iosif Shklowsky. It occurs in a particle accelerator called an electron synchrotron, a machine used to accelerate electrons to extremely high speeds. Held in an evacuated circular pipe by powerful magnets, the electrons repeatedly pass through zones where more energy is pumped into them, pushing them ever closer to the speed of light, so that the magnetic fields in the curved sections also have to be boosted to hold the increasingly energetic particles on course within the pipe. Because electrons are lighter than other charged particles, they can be made to go faster, closer to the speed of light, for a given energy input. Unfortunately for the experimenters for whom that speed (or the increased mass associated with relativistic speeds) is their main objective, acceleration itself becomes an increasing cause of energy loss – and in this arrangement even 'coasting' electrons are accelerating: under the influence of the big magnets their path is bent, which means that they are accelerated towards the centre of the ring, otherwise they would hit the walls of their container and be lost. As a result, more and more of the energy that is pumped in is converted to radiation – synchrotron radiation, it is called. In fact, in recent years, Britain's principal electron accelerator has been adapted to take advantage of this parasitic effect: instead of experimenting with the high-speed particles it uses them solely as a powerful source of this radiation.

Studies of synchrotron radiation on Earth showed that it had the same characteristics as the glow from the amorphous region of the Nebula. So: could some kind of synchrotron mechanism be found in the Crab? This would presumably require the interaction of a magnetic field and electrons travelling at relativistic speeds. It would also require a powerhouse to accelerate electrons almost to the speed of light, to be a continuing source for all that excess energy emerging as radio. How could all that come about? Theories about what happens when a star runs out of fuel were reexamined.

When Stars Collapse

Our own Sun, a modest-sized star in sedate middle age, may exhaust its fires in another 5000 million years or so. At the end, its internal radiation pressure will fail and theorists calculate that it will collapse in on itself, packing the atoms much more tightly. As its mass falls 'downhill' towards the gravitational centre, its potential energy will be converted to heat. The Sun will become a dwarf star, about the size of Earth, and will glow white hot before it gradually cools to final obscurity. Many such white dwarfs have been seen by astronomers.

But this is not the only possible outcome of such a collapse. For example, if the orginal star is many times more massive than the Sun, the gravitational forces released would be greater than any repulsive forces known to exist inside atoms. In other words, there is no force known to physics which could withstand indefinite collapse. Further, if any unknown force did exist it would lie beyond our capacity to observe it in action, because at an early stage in this process the star would collapse beyond the point at which any kind of radiation, any kind of information about what was happening, could climb out of the gravitational hole – a 'black hole' – into which it had sunk.

But a black hole was not the only theoretical alternative to a white dwarf. There is a range of masses between a lower threshold estimated to be several times the mass of the Sun and an upper limit of about as much again, within which a different kind of collapse might take place. In this, the gravitational force is great enough to overcome the electrical force which gives atoms their electron clouds and is the basis of all of their chemistry. But it would not be quite strong enough to overcome the forces which act inside the nucleus itself. As a result, atoms collapse into themselves; electrons are absorbed into their nuclei, or into close range around them.

If a proton (the positively charged nucleus of a hydrogen atom or a unit in the construction of other nuclei) is made to combine with an electron (carrying a negative charge) the two electrical charges are balanced, or cancelled, and the result is a neutron, a neutral particle with a mass slightly greater than that of the proton. Accordingly, a star of this kind was called a neutron star by the theoreticians who predicted it. It would be tiny, about ten kilometres across for a collapsed star with the mass of the Sun, and it would have extraordinary properties due to the intense gravitational field at its surface. Says Mal Ruderman, a New York theoretician:

> If we were to move on the surface of a neutron star, our heads would weigh as much as a hundred large ocean liners. The strength of the surface would support structures – let's call them mountains – a foot or so high. It would be an exotic picture, because the atmosphere would be only a few inches deep and the mountains would stick right through. It would be an interesting sport to try to climb one: it would take the energy expenditure of a whole lifetime to move up an inch!

This was the kind of star which a few more adventurous physicists suggested might be at the heart of the Crab. But how would it supply the power? And how on earth (or from Earth) could an object only a few miles across be detected at a distance of 30 000 million million miles? The neutron star was regarded as 'a mythical beast', of no practical interest to astronomers who had better things to do with their time and their telescopes . . . until 1967, and a discovery that at first had nothing to do with the Crab.

Little Green Men?

A new radio detector, built in 1967 at Cambridge, was perhaps the least glamorous telescope ever constructed: it looked like four and a half acres of washing lines, an array of over two thousand separate antennae. It was actually built to search for quasars, objects presumed to be at enormous distances across the Universe but, unlike galaxies, so compact that to the optical astronomer they would twinkle like stars when seen through the shifting air masses of our own atmosphere.

The new radio detector was designed to scan the sky as Earth rotated beneath it. Different tracks across the sky could be followed by combining the signals from the parallel rows in special ways. However, its most important characteristic was that with such a large area to collect the signal, this did not have to be accumulated over a period of time; instead it could be recorded in real time, in fact, as a pen trace showing the strength of radio noise in the beam from moment to moment. Under the supervision of Anthony Hewish, his graduate student, Jocelyn Bell (now Bell-Burnell), had helped to build the array and then went on to operate it. As she recalls the experiment:

> After we'd been running a few months, I began to notice something curious in the records. Of course, on these charts you could see radio sources and you could also see manmade interference. But there was also something that didn't quite fit either bill: it wasn't exactly a twinkling radio source, and it wasn't exactly interference either. It was coming at the same sidereal time each day; in other words, it kept pace with the stars, not with any of our manmade activities, which go by the Sun or Greenwich Mean Time.

As though to make matters more difficult, the signal was weaker at each pass but, after carefully going over three and a half miles of the paper charts, Bell decided that it really did look like a star that was ticking like a clock. Sharp spikes in the trace (when the signal was good enough for them to stand out clearly) were marking time in precisely regular periods of just over a second. Hewish's initial reactions were of extreme caution, the only proper response, he felt, to something which defied belief. Since it was outside all previous experience he maintained for as long as possible his scepticism in the signal's natural origin. This was a general response, as Jocelyn Bell admits:

> Everybody's first reactions were that it must be manmade. The second reactions, not really voiced very loud, were 'perhaps it's little green men', another civilisation. They could

> produce a manmade signal; they would be close to the star, like our Sun; and they'd move round with the stars. I think, perhaps, being younger myself, I found it slightly easier to believe than some of the older hands at the game. But we did really spend a lot of time with that first one, trying to explain it away. We wrote to all the observatories in Britain, saying, 'Have you had any programme going, which could cause radio interference?' Because, of course, the only people who keep to sidereal time are astronomers.
>
> It was easier with the second one, and that was a great relief, because it removed the possibility of 'little green men'. It was highly unlikely that several lots of little green men would all be signalling to us at the same frequency, all at the same time.

Delays in publishing due to their caution were widely misunderstood outside Cambridge. News of the first pulsing star was not published until the second was found, and the co-ordinates of that were not disclosed until later still. Was Cambridge hanging on to essential details in order to make their own killing before anyone else could get a look in? Also, with traditional Quaker modesty and patience, Bell did not promote her own cause and many, especially in America, already irritated by what they saw as Cambridge's secrecy, were inclined to believe that she had received too little credit for her persistence. For herself, Jocelyn Bell simply felt that she had been lucky to have been given the opportunity, and to have been in the right place at the right time, and that all else followed. In due course, she wrote her thesis, in which one of astronomy's most exciting discoveries was relegated almost in passing to an appendix. Her name does not figure in the Nobel Prize that was later given for early work on pulsars, as they had come to be called, but such anomalies are standard Nobel procedure.

Pulsars

The discovery of pulsars began a kind of Galactic goldrush, reminiscent of the hunt for quasars a few years back or the activities of the comet hunters of two centuries before. Once astronomers knew what to look for and how to find them, new examples poured in. Typically, they would pulse, each with its own distinctive rhythm, at intervals of around one to three seconds. The majority were in or near the plane of our own galaxy and so were likely to be part of it: but what were they?

Naturally, astronomers tried to explain them in terms of stars they already knew something about: a white dwarf was the popular candidate. This was, it appeared, of such a size and density that, if it could be made to vibrate, it would oscillate quite fast – in fact, with a period of a few seconds or more, fairly close to the observed pulsations. At a pinch, a period as low as one second seemed possible. But then a pulsar was found in the Vela Nebula that had a period of a tenth of a second. The white dwarf hypothesis had begun to look sick when the final blow to it came from the Crab.

Some of the earlier pulsars had already been discovered by the National Radio Astronomy Observatory at Greenbank in West Virginia. In a fold of the Allegheny Mountains lay a group of instruments that included a 300 foot semi-steerable dish, and this was turned, hopefully, to the Crab. The technique had worked well enough for pulsars elsewhere in the sky, but for once the Crab did not respond. No blip appeared on the trace. Staelin and Reifenstein, the two observers, passed the signal through a computer that was supposed to separate any signal from the background noise – and still there was no pulse.

However, the telescope was also set up to receive and analyse a whole range of radio wavelengths, in order to measure a property of the pulses that was called dispersion. Ordinarily a pulse is heard as a sharp single click, but that is true only for an observer situated in the same region of the galaxy. As it moves through the interstellar medium, the high frequency components of the pulse move a little faster than the low, so that to a listener who is thousands of light years away, the click is replaced by a descending whistle. The rate at which the whistle descends is, in fact, a good indication of distance. But if the slides are so close together that they overlap in time, they disappear from the trace, especially if there is a great deal of other radio noise.

Staelin and Reifenstein took data for three weeks and then buried themselves in a computer room with their printouts, huge piles of paper with many numbers across each sheet, and then down the pages, fold after fold after fold. . . . If there were a pattern, it was peculiar and difficult to see, though there did seem to be some evidence that the highest numbers were related along the same diagonal angles. Was there a computer problem, or perhaps some kind of interference with the signal?

'We then spent a whole weekend on our stomachs,' said David Staelin, 'over a large amount of detailed printout giving every

The 1000-ft radio dish in a natural hollow among hills at Arecibo, Puerto Rico. The telescope observes a particular object by moving the antenna to the point above the dish at which its image is focused. The pulsar in the Crab was discovered here.

number we had received, and attempting to find some pattern from this plethora of numbers. So we circled the large numbers and, out of this seeming randomness, the pattern we were searching for emerged, a descending whistle. We knew we had a pulsar.'

But with only a few whistles emerging clearly from the noise, and those at erratic intervals, they could not work out the period. Their data and method were sent to Arecibo in Puerto Rico, which with its 1000 foot collecting bowl had over ten times Greenbank's sensitivity and was well placed to pick up the Crab. A team at Arecibo quickly confirmed the pulsar and soon had the period too. It was faster than any previously known, flashing at a rate of thirty pulses every second. Even more startling, within a few days it became clear that the period was changing. It was slowing down.

The Energy Source

Arecibo is an out-station of Cornell University in up-state New

York, so the news of the slowdown was transmitted by radio direct to the Cornell Space Sciences building, where it was of great interest to Tommy Gold. Gold was a theoretician with a reputation for ideas that were brilliant but sometimes a shade too innovative for his fellow scientists. He was a co-author with Fred Hoyle and Hermann Bondi (both since knighted in Britain) of the Steady State Theory of cosmology, an elegant and tidy way of arranging the Universe, but which by then seemed increasingly out of line with the evidence. Later he had stood up for neutron stars as the most likely nature of the pulsar but, as he saw it, he had virtually been told to sit down and shut up by astronomers who were happy with the pulsing white dwarf hypothesis.

When Gold was handed the information on the spin-down of the Crab, he saw that its clock was slowing by about three seconds each day, about a thousand seconds in a year. The implications were obvious to him: once he knew the slow-down for the pulsar, he could work out how much energy was being lost. Familiar with the mass and size of neutron stars, Gold made a hurried calculation which produced an answer of 2×10^{38} ergs per second. He then needed to compare this with the luminosity of the Crab – the total requirement for the mysterious energy source. Gold recalled that, some years before, the Russian, Shklowsky, had made an estimate for that, and immediately sent someone to the library for the article.

When this was found, they saw that Shklowsky had written that relativistic electrons were being injected into the amorphous region at a rate of about 10^{38} ergs per second, incredibly close to the figure Gold had written down. Plainly they had the answer. The energy being lost from the pulsar was much the same as the energy with which the rest of the Crab was glowing at radio frequencies; this was, indeed, the mysterious energy source of the Crab.

But they had gained more still: in order to get their figures for mass and size, they had had to take into consideration that it was a neutron star. So Gold could claim that at a stroke it had been shown that the object in the Crab, the pulsar, was indeed a rotating neutron star, deriving its energy from its spin and was flashing to us like a lighthouse beacon.

The hiccuping white dwarf was dead – to be supplanted by a far more exotic object spinning at thirty rotations per second. And fantastic though that rate might be, it was easy to calculate how an even faster rate (as it must have been to start with) could have come about. If the original star was rotating even slightly to begin with when it collapsed to the size of a neutron star, a simple physical principle, the conservation of angular momentum, would come into operation so that the speed of rotation would go up dramatically – just like a pirouetting ice skater drawing in his arms and legs to increase his rate of spin, but on a much greater scale.

Not only that, but the collapse would have increased the intensity of the star's magnetic field to perhaps a million million times that of Earth. At Princeton, Jerry Ostriker calculated what would happen. He imagined a powerful bar magnet with its field extending far out into space. He then supposed it to be spinning with the north and south ends rotating at high speed about the middle. The magnetic field would be carried round with it. At only 10 000 miles out from the star, a short distance on the scale of the whole Nebula, a point at the end of a line drawn straight out from the surface of the star would be moving at the speed of light. By the principle of relativity it is assumed that no object can exceed the speed of light and Ostriker, applying this also to the field lines from his bar magnet, found them to bend back into a spiral. Spinning at high speed, this spiral looked like a wave progressing outward from the neutron star, and a charged particle such as an electron could ride on that wave.

This was all that was needed, not only for the quantity of energy required but also for its supply in a form that could accelerate the electrons directly outward until in the amorphous region they would be travelling at close to the speed of light. Then they would radiate in just the way that the Crab is seen to do. They would continue to receive energy from the field-waves and to emit it again, as far out as the field had sufficient strength.

Furthermore, some of these electrons and their radiation would strike matter in their path and push those ever faster outward too. So the original expansion might not slow down, as would be expected in a dense interstellar medium, but could actually accelerate as time went by. This would account for the discrepancy between the presently observed expansion rate, which points back to an explosion that seemed to have taken place around AD 1150, and the actual sighting by the Chinese, which was about a century earlier. So it now seemed that many of the big questions about the Crab, including its brightness, appeared to be solved: however, the source of the pulses was still a mystery.

A Tale of Two Tapes

Optical astronomers had already begun their own search for pulsars, stars flashing with visible light, in the hope that the extra information might provide some clue about how the pulses are produced. Because nobody had ever reported seeing flashing stars, they expected that any optical effect would be faint. So, working with a small telescope in the grounds of the Astronomy Department in Cambridge, Roderick Willstrop used neither his eye nor photographic plates, which would certainly smooth out any pulsation; instead he placed a photomultiplier at the focus of his telescope. This amplifies the light that falls on it at any moment, so that any fluctuations can be recorded. In an ideal case, such variations would appear on a pen-trace as clear peaks and troughs, but in practice it was a great deal more likely that the blips would be small compared with the effect of our own atmosphere on other, steady light picked up at the same time. So it was sensible to record the signal on punched tape for subsequent analysis by computer.

Willstrop has good reason to remember one particular night spent on a small telescope in the gardens of his Institute. At Arecibo, unknown to him, they were still convincing themselves of the spin-down of the Crab; Gold, almost at that very moment, was proving to his own satisfaction that pulsars must be neutron stars; but Willstrop, who had been patiently searching for optical pulsars for some time, did not have this vital extra clue. Although, like everybody else, he was still more interested in white dwarfs, some pulses had recently been reported from the region of the Crab, so it was certainly worth taking readings there, too.

Willstrop's record of the night of 23–24 November 1968, is dominated by the comings and goings of clouds. He went to bed at 21.30 because it had clouded over, and got up again at 22.30 when it cleared and he decided to observe the Crab. He found it easily enough: the sky was clear enough for him to record that 'starlike condensations' were visible and he thought that if he could find a photograph he might identify the central object. He went to a library at 23.55 but when he came out again at 00.10 it was cloudy. Finally at 02.30, it began clearing from the west: at 02.56·5 his notebook records, he began the vital run, which lasted for 11.5 minutes. But Willstrop's observational technique was better than the programme worked out for his computer, so the results were not analysed at once, but simply stacked up with other paper tapes to await processing.

Soon after that night, two Americans, Don Taylor and John Cocke, were joined by an Englishman, Mike Disney, on Kitt Peak, a mountain in Arizona. They had just heard the news from Arecibo, giving the period of the pulsar, and had decided to take a look at the Crab themselves. In fact they used, not one of the big telescopes for which Kitt Peak is world famous, but a relatively small instrument belonging to the University of Arizona's Steward Observatory. None of them had much experience of practical astronomy, but with the help of the night assistant they found the Crab and held it with some difficulty. The night was cold and clear, but windy, and the wind buffeted the telescope.

Knowing the exact period, they needed no computer: any pulse could be displayed visually by chopping the signal from a photomultiplier into segments of the right duration and adding them up. They also recorded their observations on a quarter-inch tape, identifying each run by plugging a microphone across the circuit and speaking into it. The crucial observation begins with Cocke's announcement. 'This next observation will be observation number eighteen,' after which the microphone should have been disconnected. But, in the event, the plug was not pulled out far enough, so the tape recorded not only the hiss and crackle of the Crab but also the voices of its observers. It gives a fair example of the reactions of astronomers at a moment of triumph. After a pause, over the noise is heard, in tones of increasing astonishment:

'We've got a bleeding pulse here!'

'It can't be.'

'God!'

'It's a bloody good pulse. Just come and look at it down here.'

'This is a historic moment.'

'Christ Almighty, look at that!'

'God! I really can't believe it, you know.'

Willstrop heard of the discovery about eight weeks after his own observations:

> My first reaction was of great disappointment. Of course, in these circumstances, the most important thing to do is to pick up the pieces and to use the observations to confirm the discovery. . . . If you're going to be first with results in scientific research, you have to have the luck to make the right decisions, and to analyse the right data. And in this case I didn't.

The Pulsar Lighthouse

If a recording for the pulse from the Crab is replayed and shown on a video display, repeating itself thirty times per second, and the signal is added cycle by cycle, it gradually forms itself into a clear line. Paul Horowitz, with his friends Papaliolios and Carleton at Harvard, interpreted a trace built up from about twenty thousand cycles (some ten minutes of data). It shows a main peak that has a very sharp tip. This indicates that, not only are the individual pulses very sharp, they must also be lined up to better than one ten-thousandth of a second. From this measurement alone, Horowitz calculated an upper limit for the size of the radiating region of the pulsar. It could not be larger than about thirty kilometres across, and possibly less of course. This was a direct observation, confirming the calculations of the theorists and it showed that the electromagnetic (light and radio) pulses must be coming from a very confined space in or near the star.

The trace showed another feature: a second bump, broader and flatter than the first, half a cycle later. In fact the trace was consistent with the picture of a star spinning at the fantastic rate of over thirty revolutions per second, emitting two distinct lighthouse beams from opposite sides; the two beams being imperfectly aligned, or otherwise slightly different from each other.

How could such a lighthouse be arranged? A plausible suggestion is that the radiation emerges in line with the two magnetic poles, where there could be a very rapid escape of charged particles such as electrons or protons along the magnetic axis. This might create the beams of light and radio waves. Then, as the neutron star spins, we might see a flash each time it points towards us. This also implies that we must be in a favoured plane relative to any pulsar from which we see the pulses, and that we are lucky to see the Crab pulsar at all.

The next stage was to identify which of the stars in the Crab was flashing in this way, just too fast for the eye to see. Years before, Rudolph Minkowski had been observing the Crab by eye, using one of the big Hale telescopes, and a woman with the team had queried whether a particular star was flickering very fast. But none of the men could see it and, by the end of the night, they had all discounted it. Later, Minkowski wondered whether there really were differences in the flicker rates which men and women can perceive and whether, with a few more female astronomers around, the optical pulsar might have been recorded earlier.

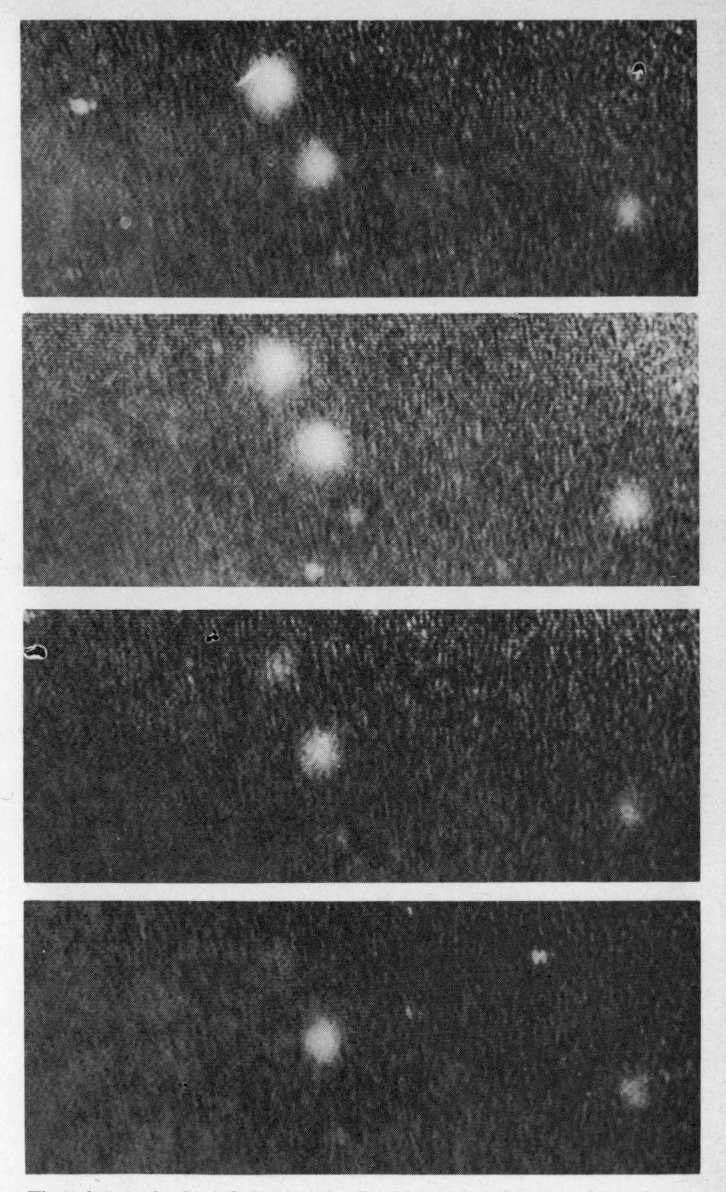

The pulsar in the Crab flickers too fast for the eye to see, but it can be made to appear and disappear – and so be identified – by means of a stroboscopic shutter.

Using the 120 inch telescope at Lick Observatory, Joe Wampler devised an ingenious way to make the actual star very easy to identify. He rigged up a makeshift spectrosope at the focus. It would produce an effect similar to that sometimes seen in Western films, in which wagon wheels seem to rotate backwards or stand still at certain speeds. This is due to the intermittent action of the film shutter, producing the effect when the frame repetition rate is close to that at which successive spokes are in the same position in each new frame.

Wampler used a disc with radial slots and looked through it at the Crab. When he rotated the disc at speeds close to the known pulse rate, one particular star could be seen to flash slowly or, with a better match, could be turned on or off at will. From the moment he switched it on, the pulsar was obvious. Wampler recalls that his friends told him that he had 'jumped up and down a little bit'. He himself did not remember doing any jumping, but neither did he feel able to deny the rumour. Wampler was the first to know that he was actually looking at a neutron star, the powerhouse of the Crab.

The Glitch

Timing the slowdown of the pulsar's spin had become a game in which anyone with access to a modest sized telescope and an aptitude for electronics could join. Arecibo had continued to record the radio signal (for which a big telescope was still required) but, in the summer of 1969, Arecibo had technical problems and others had lost interest, leaving Ed Groth, a graduate student at Princeton, as the only observer when, in late September, the pulsar suddenly went haywire. The pulsar clock seemed to jump, so that it was running a little faster. They called it a 'glitch.'

The spin-down had been easy enough to explain as a result of tidal drag due to energy loss . . . but how could they account for spin-up – an increase in speed? The Princeton data were thin: the rate of change rested on a single point on a graph. If it was wrongly placed the effect might be due to some kind of slow oscillation, rather than a sudden jump. The one-point observation was not published at the time and there was scepticism about whether the event had really happened. But two years later, Nelson and Hills, two students using a 24 inch telescope at Lick, saw another glitch in the Crab pulsar.

According to Mel Ruderman, there had been no shortage of

theories about the detailed structure of neutron stars. He would expect that the surface of the pulsar had solidified soon after it formed and that initially it had a different shape, bulging more at the equator as it rotated very fast. Then, as it slowed, the crust would be strained as it tended to settle into a more nearly spherical shape.

'What we don't know,' said Ruderman, 'is how the crust manages to relax and get rid of the strain. It could flow plastically, like molasses. Or it might crumble like a cookie. Or it might snap now and again, like earthquakes on Earth.'

However, only in the last case would there be a change in the timing such as had been seen. So the glitches were evidence for starquakes in a solid crust. Whether that was the explanation or not, there has been an additional effect: following the 1969 glitch, fuzzy wisps of material near the pulsar appeared to move violently away from the neutron star, afterwards returning to their original position as part of a continuing ballet of motion in the amorphous region of the Nebula.

Windows on the Crab

Light from the Crab has been studied in precise detail for eighty years and radio waves for thirty, for one simple reason: light and radio waves reach the ground because Earth's atmosphere has open windows in those electromagnetic ranges. So recent work on the Crab has concentrated on opening other windows, to observe its infrared, ultraviolet, x-rays and gamma rays.

One advance was made in the early 1960s, when x-ray rocket detectors were first sent up to observe the Crab. The x-ray counter was shielded on all sides but one, on which it had a perforated honeycomb of metal to allow in only those rays that travelled along its channels. Through the defocused eye of this detector no detail could be seen: but the power it recorded was fantastic – there was far more than in the Crab's light and radio combined. Then, when an x-ray rocket was launched just as the moon was about to pass in front of the Crab, it was found that the x-ray source disappeared slowly: it was an extended source matching the amorphous region, not a small one like a star.

The results from the Crab provided a powerful stimulus to x-ray satellite astronomy, for the development of imaging cameras and real-time counters, which have paid off with startling results in many other areas of astronomy besides the Crab. Other satellites have demonstrated its ultraviolet radiation (present but not spectacular), and its gamma rays. Balloon flights carrying detectors for very high energy rays have confirmed its range of wavelengths and power.

The Crab has never disappointed its observers: it seems that whatever the experiment the Crab is there waiting to show up, nearly always among the first objects to be seen, and often among the brightest in the sky, so that within a few years of the pulsar's discovery it could be calculated that it was powering an output of energy 25 000 times greater than our Sun. If it is like that as we see it now, what can it have been like on that day over nine hundred years before it first appeared in our skies, brilliant even by day? That incredible violence might well account for the formation of an abundance of elements that are heavier than iron. But to confirm that we would need to train out instruments upon such an inferno on the day that it occurs. There are secrets still to uncover.

The Crab is unique only in the time and place of its appearance, which have made it so rewarding an object of study; it is not unique as an event in the recorded history of our galaxy, just rather rare. Several other supernovae have been seen since then and more are likely to have occurred in regions hidden from us. There is a variety of supernova remnants, still expanding or pressing against outer regions of interstellar gas. Such an event, about 4600 million years ago, in our own part of space may have triggered the formation of the solar system and seeded it with most of the elements of which the bones and flesh and blood of our own bodies are made.

The human brain is made up of elements that were not created here on Earth but were whizzing about in space before the solar system was formed. Astronomy is the science by which those very atoms and molecules now crammed inside the human head try and find out where they came from. Until the middle of the twentieth century, they have been able to collect information only through the eyes, through the visible spectrum of light.

If we are to continue our imaginary journey beyond our own galaxy we shall need more information than that narrow optic window can provide. The Crab Nebula, sharing experiences with us in this swirling disc of stellar material we call the Milky Way, is relatively close and easy to see. It is only 6000 light years away. Our neighbouring galaxy, Andromeda, is two million light years away. Travelling across that vast space, visible light becomes diffused and only a small amount reaches us. Although Andromeda is as big as the Moon in our sky, it can only be seen as a faint blur, hardly brighter than the black background of the night. Anything further away vanishes into the darkness. Optical telescopes cannot increase the amount of light reaching us. If it was not for the fact that stellar objects release phenomenal amounts of energy outside the visible spectrum, that might be the end of our journey.

Fortunately, those atoms inside our brains have devised a new branch of astronomy, which can measure and chart all the very high energy, very short wavelength radiation that, if we could see it, would make the sky glow. We can continue our journy into the darkness, bathed in a new light.

7 Darkness Visible

Dick Gilling

The Second World War did not bring many benefits to the human race in general but, as this planet darkened, the remainder of the Universe became more visible. One very simple result of the war was the blackout of lighting over the urban sprawl of California. Intended to conceal possible targets from Japanese bombers, the sudden dark quite incidentally made possible nearly perfect conditions for the great telescopes on Mount Wilson and Mount Palomar. Optical astronomers depend on clear skies and dark nights: clouds hide the stars and the scatter of street lights fogs the film in the cameras sufficiently to obscure tiny points of light far out in space. So the blackout was something of a golden age for any observers not otherwise engaged during the war years.

Rather less directly, the development of radar opened a new window on to the sky. Powerful electronic equipment and sensitive antennae had been used to detect the feeble radio waves reflected by aircraft. Without radar, the Battle of Britain would probably have been won by Germany and radio astronomy would probably not yet exist. As it was, a few years after the war ended came the amazing discoveries of the radio astronomers, who were now able to picture whole galaxies which the optical astronomers, even in the warm darkness of wartime California, had not been able to see.

A third discovery came from the Second World War by a yet more circuitous route. While we can detect by eye the part of the spectrum of electromagnetic waves which we call visible light, other parts of the spectrum are hidden. Infrared, waves of a lower frequency than visible light, and ultraviolet, higher frequency waves, are both familiar: we have all seen infrared photographs, sometimes called heat pictures, and Mediterranean beaches are hidden from view every summer by human skins burning under the same ultraviolet radiation that lights up white shirts in the discos. At a yet higher frequency than ultraviolet in the spectrum come x-rays. If x-rays could reach the sun-worshippers, skin would be not merely bronzed but toasted to a crisp; but by a fortunate provision of Nature, Earth's atmosphere prevents the rays from reaching the surface of our planet. The wartime invention that made this new form of astronomy possible was the first device to go above the atmosphere – the V2 rocket.

X-ray astronomy, for that is the subject of this chapter, is thus a new branch of what must be the oldest science. Its pioneers, and to a great extent its practitioners today, began as physicists. Since their discovery by Roentgen in 1895 – as an emission from an evacuated tube – physicists have been fascinated by the origins of x-rays and the states of matter that they signal – x signifies the unknown, after all. X-rays are emitted by matter in a hugely energetic state, very hot matter, for instance. The nearest very hot object in the sky is the Sun and it was, naturally enough, the Sun which was first examined for x-ray emissions.

Rocketry

If any one person could be dignified as the Father of X-ray Astronomy, it would be Herbert Friedman of the United States Naval Research Laboratory (NRL). Friedman, now in his sixties, had worked on radio communications for the US Navy during the war. He, with many others, had noticed interference with Navy radios during periods of activity on the Sun and had suspected that it was due to interference with the ionosphere, the electrically charged layer above our atmosphere. Perhaps the interference, thought Friedman, was due to x-rays from solar flares. If so, would it be possible to detect such exotic things?

By one of those happy accidents that mark the progress of scientific achievements, the US Navy had liberated from Germany a number of V2 rockets which Dr Werhner von Braun had failed to send on their customary journey to London. The Navy brought them to the White Sands Missile Range in America for testing under more peaceful circumstances. The V2s were not in good condition. On their transatlantic journey they had been exposed to corrosive salt spray and in the New Mexico desert to sandblasting; but from this unpromising selection of secondhand missiles enough spare parts remained to rig up a few functional rockets which would be capable of lifting instruments above the x-ray opaque atmosphere. It was too good an opportunity for Friedman to miss. He and his team tried first a simple device: photographic film, covered with metal foil, was taped to the outside skin of the rocket. No doubt x-rays penetrated the metal foil as intended, and perhaps produced an effect on the film, but the film packets were too fragile to stand the return to Earth. It was not until Geiger counters, suitably shielded, were launched on the German rockets that any convincing evidence of the existence of solar x-rays was returned to Earth. That was in 1948.

The Sun emits only a millionth part of its energy in the form of x-rays. But because it is so near to us, the rather primitive instruments of those pioneering days were able to detect the radiation. Friedman and his group, using improved Geiger counters, went on to monitor the Sun's x-ray output over an entire eleven-year sunspot cycle. At the same time, a group in Britain, led by Robert Boyd at University College, London, had also begun to experiment. They were using rockets launched from the Woomera Range in Australia. Both groups now look back to those salad days with affection. They were plagued with technical problems, sometimes because rockets exploded prematurely or because the electronics, in the age of glowing valves, misbehaved. Always there was the problem that observations could be made over a period of minutes only, as the rocket reached the top of its trajectory; and there was always tension in the period before the instruments fell to Earth for recovery. But the x-ray groups were a special category in the astronomical sciences: the brotherhood of optical astronomers either ignored them or took them with less than total seriousness. They were, after all, only looking at the nearest star, and they were physicists, not really astronomers at all – astronomers look at all the stars.

It had, of course, occurred to more than one of these errant physicists that if our nearest star, the Sun, was an x-ray emitter, then more distant and interesting stars might be doing the same thing. Indeed, in 1956, in the course of detecting solar flares, Friedman's group had encountered some emissions in the higher frequencies of the x-ray spectrum (there is a spectrum of x-rays just as there is of visible light) which powerfully suggested an x-ray source that was not the Sun. The results were reported to a meeting of the International Astronomical Union in 1958, in Moscow and reverberated around the astronomical world like the sound of a penny falling in mud. Full acceptance of the x-ray astronomers would have to wait.

At just about this time occurred a further circumstance which looked like a curse, but turned out to be a blessing. While the NRL and University College groups, with scattered help from elsewhere, were making observations of the Sun in x-rays, a new branch had started to grow in New England. Bruno Rossi, at the Massachusetts Institute of Technology, was on the panel of advisers for the US space programme. A fellow Italian, Riccardo Giacconi, a young physicist who had been studying cosmic rays, was waiting for clearance from the American Government to work on a space contract. Rossi mentioned the possibility that something big might be building up in the new field of x-ray astronomy and Giacconi, during his six-month wait for clearance, began to think about the problems and the possibilities. The Sun was no longer interesting, since plenty of work was being done on that already: the stars were a different matter.

Hardware and Funding

The first problem Giacconi saw was in the instrumentation: existing instruments were simply not sensitive enough to detect the tiny x-ray signals that were expected. With all the revolutionary fervour of a newcomer, Giacconi decided, with his colleagues, to make a radical philosophical change: rather than seek accuracy by searching a small area of sky, his idea suggested a wider angle of search, making the statistical chances of finding an x-ray source much better.

The sensitivity of the instruments could be increased by making them much larger, which created a problem only insofar as it would increase the payload of the rocket. Even better sensitivity would result from a device to somehow eliminate interference from cosmic rays, which would otherwise confuse the x-ray statistics; this could be done by installing a cosmic ray counter and simply substracting the cosmic ray count from the x-ray count. With these three ideas, Giacconi and his collaborators managed to increase the sensitivity of the old Geiger counter style instruments a hundred times. That was possible. Giacconi had in mind another device which most people thought impossible, and Rossi advised him to delay work on this until results from the conventional instruments were convincing enough to attract funding for something more radical.

By the late 1950s funding was already far less easy to come by than it had been in the relatively carefree days of the V2 when Friedman and others had made the first observations. Scientists who needed large sums of money had to develop the skills of entrepreneurs and consider that final arbiter, the bottom line – who would profit by their research and by how much? Any straightforward commercial profit was profoundly unlikely, so arguments had to appeal to the US Government agencies. Scientific or science prestige value might attract NASA; the only other likely benefactor would be one or other of the military arms: Army, Navy or Air Force. Which would be the most likely to support any further progress in x-ray astronomy, even when its most sanguine adherents would admit that practical results (or any results at all) were rather unlikely.

Giacconi put his ideas on paper in the form of a proposal to NASA, only to have them turned down. Ostensibly the reason given was a lack of confidence in x-ray astronomy as such, but in fact NASA was already supporting two other groups (neither of them ultimately successful) and saw no point in subsidising a third. This created a problem: clearly NASA was being less than frank with its applicants. Two can play at that game, and frequently do, so Giacconi and his colleagues decided to approach the US Air Force in order to get the money. At that time the Air Force was interested in the Moon and so they therefore suggested that perhaps the Moon could be detected as an x-ray emitter.

It was at that time that the word 'Moon' was an almost gilt-edged guarantee of funding, as the word 'environment' would later become. The submission to the Air Force does in fact also suggest that the instrument should look at the stars as well as at the Moon: Giacconi's detractors would later suggest that only the Moon was on his mind, but they underrate his subtlety.

Early Discoveries

The experiment flew in June 1962. By a quite amazing piece of good luck, its instrument pointed at a star in the Scorpius constellation whose output in x-rays was a thousand times its output in the visible spectrum – almost the exact reverse of the Sun.

Such beginner's luck is surprising and it was natural that the small community of x-ray astronomers should receive it with some disbelief. Scientists, like the British pioneer Ken Pounds, wanted confirmation. Within a year both Dr Giaconni's group and Dr Friedman's team at NRL had provided this and went on to discover a fresh source in the Taurus Constellation. This was later associated with the Crab Nebula.

With these discoveries, the x-ray astronomers had by 1963

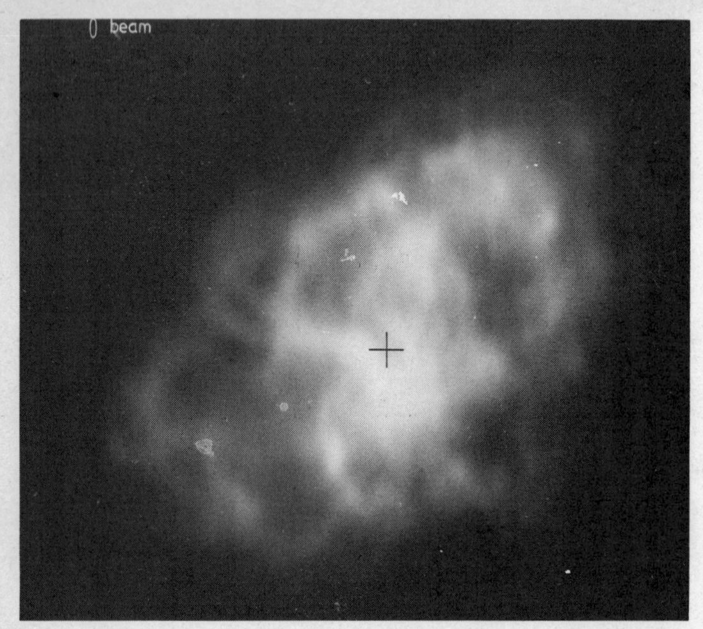

The Crab Nebula as the human eye cannot see it. Above: *as a radio 'picture' from the Cambridge radio telescope. Although radio emissions are at a wavelength far lower than visible light, the nebula can be visualised by the Cambridge instruments.* Above right: *only just below the wavelength of visible light: the Crab Nebula in infra-red.* Right: *at a far higher wavelength, the Crab as it is seen by its x-ray emissions. The x-rays are probably generated by a neutron star at the centre of the nebula.*

proved their right to be accepted by the astronomical community. The two observations had, separately, just the right qualities to initiate a successful new discipline. Scorpius X-1, Giacconi's discovery, is a star of a very odd kind. Even now, the reason for its powerful x-radiation is not clear. On the other hand, Friedman realised early on that this discovery of the star in the Crab Nebula has theoretical implications of the most profound kind which were bound to involve astronomers and physicists in healthy discussion for a long time.

'We began, Friedman recalled, 'to think about the possibility that the x-rays from the Crab were x-ray emissions from a hot neutron star. I presented these early results in a lecture at

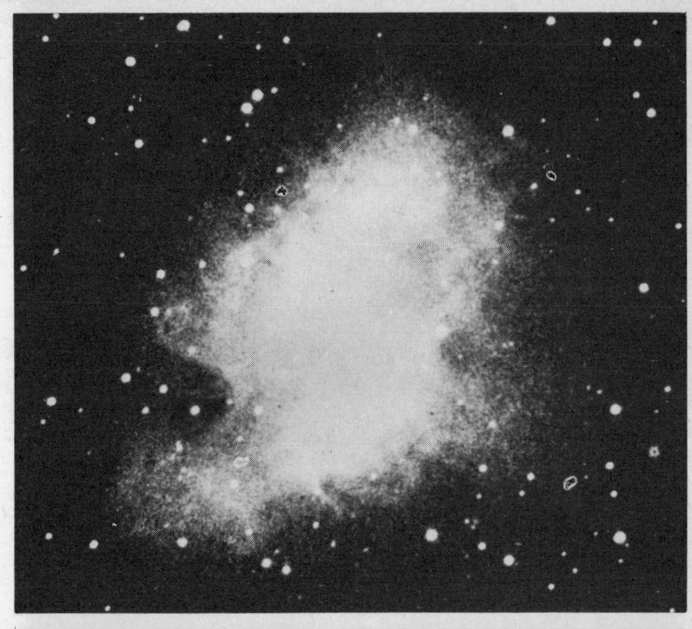

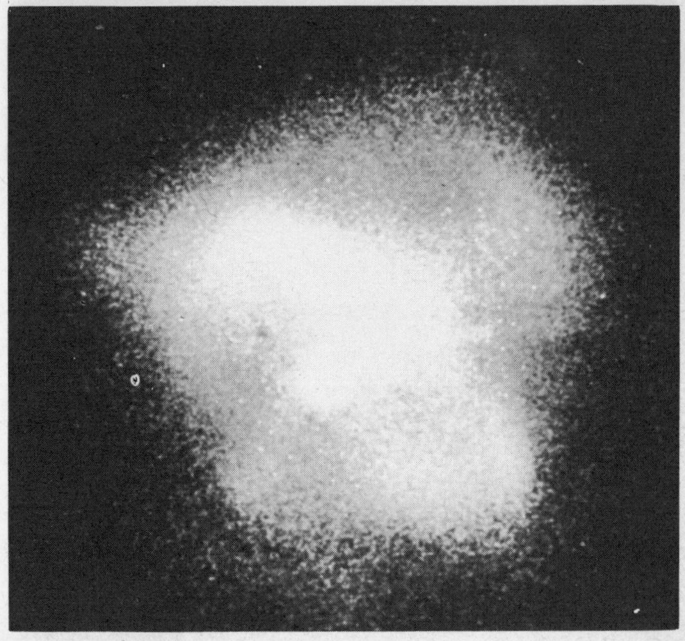

Princeton. Robert Oppenheimer was in the audience and he came up to me afterwards and remarked at how thrilled he was at the prospect that neutron stars and black holes might really be observable with the new x-ray astronomy. In the thirties he theorised about neutron stars and black holes but with the feeling that these were the kinds of things that one theorised about but could never observe.'

So one of the results would be a long running mystery and the other a step on the way towards proof of a cherished theory. It was fertile soil in which the new astronomy could grow.

By 1965, a British programme of flights by Skylark rockets was under way at Woomera to examine the sky for x-ray sources. At that time, at least, the British had a small advantage over the Americans: their funds, though perhaps smaller, were more assured. American scientific funding tends to be shorter term and more likely to be interfered with by parsimonious Senators conscious of their need to display financial hard-headedness to the electorate. American or British, the days of the rockets were full of excitement. An experiment could be designed and performed in months rather than years; there was just the right amount of healthy competition, and virtually all the discoveries were still to be made. The main problem was that each rocket flight could yield only about five minutes of observation. Even so, by 1970 with a total of only about three hours' observation, thirty x-ray sources had been discovered, most of them in our own galaxy, the Milky Way. Moreover, within a day or so of examining the data from such a discovery the British team could be reasonably sure that a telephone call to the science correspondent of *The Times* or *Guardian* would result in a few paragraphs about their work. 'It was really quite fun in those days, but still quite removed from the mainstream of astronomy,' Ken Pounds admits.

As to what the instruments on the rockets were really looking at, nobody was unreservedly certain. The data from the sky totalled about three hours and this did not give a firm basis for theorising. Thus, the consensus view was this: the spectrum of x-rays is emitted when matter is heated to very high temperatures, measured in millions of degrees centigrade; this heating may occur in several ways, but most fiercely as a result of a gravitational collapse, when the nuclear fuel of a star is used up, and it collapses upon itself to form a star of unimaginable density; if the star is large enough, it will explode, forming a supernova like the Crab Nebula and a neutron star. Even larger stars will collapse with such force that they form black holes, with a gravitational force so huge that nothing, not even light, can escape: powerful x-ray sources, therefore, are the signature of such events, the death and transfiguration of stars. But in 1970 this was mostly guesswork.

Uhuru

Despite the euphoria of the 1960s, which culminated in a grudging acceptance of the transmogrified physicists by the traditional astronomers, it was clear that the data, let alone the theories which might develop from it, were inadequate. Radio astronomers had found pulsing radio sources (pulsars) which had been located by the optical astronomers. But this sort of time variability was difficult to observe from a short rocket flight, and there were simply not enough data. One major step could solve most of these problems: a satellite.

Once again, Giacconi was quick off the mark. He would have been quicker, if the first x-ray satellite had not taken seven years from inception to launch. Those seven years, says Giacconi, included four spent working out a management scheme with NASA: but the team had time to design and redesign their instruments until they were just right. That first satellite was launched from an island off the coast of Kenya on that country's Independence Day in 1970. Because of the date, it was christened *Uhuru*, the Swahili word for 'freedom'.

There is not much doubt that the Uhuru satellite did indeed free the x-ray astronomers from their previously subservient role in the astronomical sciences. By the end of its time, Uhuru had brought the number of known x-ray sources up to a hundred, a notable achievement in itself. But far more importantly, the observations had lasted long enough for there to be at last more than a good guess as to the nature of the sources emitting the rays; the optical and radio astronomers had joined in to pinpoint the stars and make observations to help clear up some of the mysteries. Optical observations, in particular, were still more accurate and in some ways much more detailed than the x-ray observations.

Uhuru observed a number of so-called binary sources, two stars orbiting rather closely round each other. To detect such sources would, in fact, have been possible with the rockets: the first, Centaurus X-3, pulses about every five seconds, the second, Hercules X-1, about every one and a half seconds. What was not

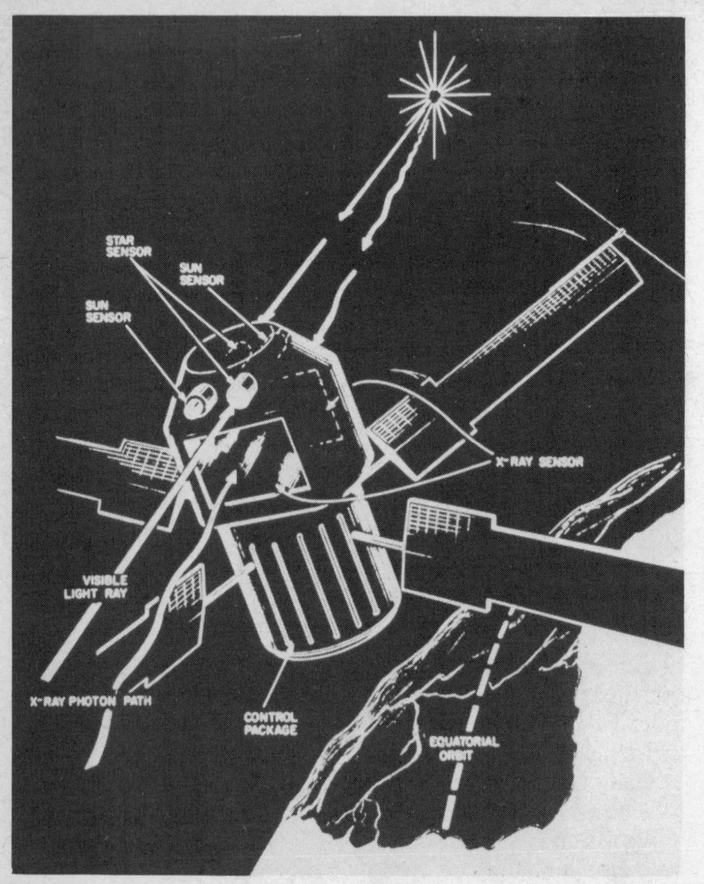

The SAS – A satellite, which became 'Uhuru' after its launch. Its powerful instruments recorded more than a hundred new x-ray sources, and gave evidence about their nature never before available.

possible with the rockets was to see that the pulses had a regular periodicity, the speed of the pulses varied very slightly, and that variation was regular and predictable. There was a matching variation in the intensity of the x-rays emitted. Put these facts together, if you are a theoretical astronomer, and you get a pretty clear answer. The x-ray star is orbiting round a companion. As it goes behind its twin, from the point of view of the satellite, the x-rays are partly masked and the intensity decreases; at the same time, because it is farther away, and light travels at a finite speed, the pulses take longer to reach the satellite and their rate appears to slow down. In time, a further surprising observation was made. The overall speed of rotation was increasing. Again, to the initiates this suggested that the x-rays were being released by gravitational energy: collapsing stars. Further calculations with the help of observations by optical astronomers very strongly suggested that the two x-ray stars were neutron stars, ripping materials from their companion stars by virtue of their extremely powerful gravitational attraction. This was the first direct measurement of the mass of a neutron star.

Black Holes

Another even more sensational Uhuru discovery concerned Cygnus X-1. This is a bright star which can in fact be seen through a pair of ordinary binoculars. Fairly early in the 1960s rocket experiments had identified Cygnus X-1 as an x-ray source: now Uhuru actually observed it make a transition. Coinciding with an abrupt decline in brightness, radio astronomers noticed an equally abrupt increase in radio emission which enabled them to locate its source very precisely. Because this transition occurred simultaneously with the x-ray transition, it was thought that they were one and the same object until photographs taken at the time were studied.

Optical astronomers then discovered that Cygnus X-1 was a binary with some optically unseen object circling round it. This object appeared to be the x-ray source and as its mass was calculated to be six times that of the Sun, it was far too massive for a neutron star. Summing up the evidence many concluded that what had been observed was an ordinary star losing matter, matter which was being captured gravitationally by something circling round the star. This was something that therefore had to be very compact and, if it was postulated that there were only two objects in that system, the only conclusion to be drawn was that that 'something' was a black hole.

So here was actual evidence of the existence of a black hole which remains the best so far discovered. Proof? Not quite. As theoretician Andy Fabian remarked: 'Nobody has yet proven that a black hole exists, and I think one of the great things x-ray astronomy has done is to make many people aware of the existence, or the possible existence of things which we might call black holes.'

To prove the existence of an object, if it can be called an

object, which is so gravitationally powerful that it does not permit even light to escape and so cannot be seen, is by definition no trivial problem. A black hole is abstract with mineral connections, a theoretical necessity whose existence can only be shown indirectly. It is not impossible that there is another explanation, but it is unlikely. As Ken Pounds points out, the theory could be useful. Most of solid-state physics relies on the quantum theory, but that theory led to the invention of transistors, and hence to microprocessors and the whole technology of the chip.

He goes on to say that in fifty or a hundred years' time it is quite possible that we may be able to exploit some of these possibilities. If the 'big bang' creation of the Universe is true there may well be a greater number of quite low mass black holes, some relatively close, perhaps even in the solar system itself. Such a mini black hole, maybe of the mass of Mount Everest, could be a very useful potential source of energy. Theoretically a spacecraft from the range of a thousand miles could project pellets of mass into it, collect the energy that came out in some large array and microwave it back to Earth.

Pounds, and everyone else in x-ray astronomy, would say, of course, that their objective is to understand more about Nature, not to exploit her. But the revelation that there are more things in the heavens than we had dreamed of certainly may in the long run lead to some practical outcome.

Satellite Observatories

The success of Uhuru led, by the natural process of scientific competition, to the launch of more satellites. During the decade from Uhuru's launch in 1970, the x-ray astronomers were harvesting a crop which had been sown in the 1960s, during a period of euphoria about space exploration and the use of space vehicles for scientific investigation that surrounded the success of the Apollo Moon exploration.

There were not very many people directly involved. Some interest was shown in Holland, Japan and in Germany, but most of the effort came from the United States and from Britain. In the US, Herbert Friedman and his group continued their research in the rambling complex of the Naval Research Laboratory in Washington. Further outside Washington is NASA's Goddard Base, named after the founding father of American rocketry, Robert Goddard. Here is the control room from which most of the scientific satellites have been monitored and given their instructions: indeed, a row of control rooms in the lower basement of one of Goddard's anonymous buildings still manages the affairs of satellites whose names are almost forgotten, but which are still about their business. In another building, a small scientific group joined in to design and install its own x-ray astronomy instruments for the satellites of the 1970s. Some hundreds of miles north and east, in Cambridge, Massachusetts, were two other groups, loosely associated with each other but hotly competitive. One was at MIT, where Bruno Rossi is now a professor, the other at American Science and Engineering Inc., with which company both Rossi and Giacconi were associated. There are other American groups, at Columbia University in New York City and in California; but the major participants are at Washington and Cambridge.

The British groups all basically stem from the original University College, London, group founded by Robert Boyd. The University College space team is stationed at the Mullard Space Science Laboratory, in Surrey. Ken Pounds who was one of Boyd's graduates is now Professor at Leicester University, and at Birmingham, Peter Willmore's Department of Space Research works on its own experiments. During the 1970s, all these groups, now in co-operation, now in friendly competition, began to push out the boundaries of the x-ray universe.

In 1974 the first British satellite devoted wholly to x-ray astronomy was launched by a NASA rocket. It was named Ariel V, the fifth in a series of British scientific satellites. It must be among the most cost-effective satellites of all time, as it was still functioning in 1979. It carried experiments from all the three main British groups and continued to add to the catalogue of x-ray sources in our own galaxy and beyond. Perhaps even more significant scientifically was a discovery by an instrument run by Len Culhane of the MSSL.

The Uhuru satellite had already established the presence of very hot gas, as hot as was necessary to emit x-rays, in clusters of galaxies. Culhane's experiment identified the gas, in a cluster of perhaps a thousand galaxies called the Perseus Cluster. Now it is possible in principle, to identify many different elements by their characteristic spectral features. In this particular case, iron appeared as very highly ionised material, that is, atoms with virtually all of their surrounding *electrons* stripped away, thus indicating the extraordinarily high temperature in the gas of something like 100 million degrees.

This one observation, then, told the astronomers of the presence of the gas, its elemental constitution, its temperature and, by implication, its origin. If it contained iron it could not be primordial gas, such as would be found at an early stage in the evolution of a galaxy, but must have been processed through a star in a supernova explosion. Ergo, x-rays are telling us about the history of the Universe as well as its geography.

Bursters

One of the American satellites, Small Astronomy Satellite No 3 (SAS 3), was not only mostly made, but also controlled by the MIT team from their own laboratories with characteristic self-confidence and professionalism – perhaps with a dash of arrogance, too. SAS 3 enlarged knowledge of neutron stars and the nuclear physics of their way of life and it made a quite new discovery of a class of x-ray sources named, with clarity if not elegance, 'bursters'. There had been such observations before, but bursters are unpredictable and they had not attracted much attention. Typically, if the term can be applied at all to such an unreliable phenomenon, a burster will flare up to emit perhaps twenty times its normal output of x-rays, then die down again: the period of quiescence can vary from a few to fifty hours. On 2 March 1976, though, SAS 3 observed for the first time a burster of a radically different kind.

Because the average interval between bursts is about fifteen seconds, the MIT team named it, with reassuring simplicity, the 'rapid burster'. The behaviour of the rapid burster, though, was by no means simple. The bursts varied in intensity; and after a strong burst, the quiet interval was long, perhaps as much as thirty seconds. A low intensity burst may be followed by another almost immediately. Since this clearly presents a problem in physics (what makes the bursts so irregular) the MIT team puzzled over possible explanations.

Walter Lewin suggested that there are instabilities, which he called 'hiccups', in the flow of the mass from the nuclear burning star to its low mass companion neutron star nearby. The fact that the material cannot flow in a steady stream is indicated by the irregular stream of x-rays, these incredible 'hiccups'. Additionally the material flowing to the neutron star may be restrained by something, perhaps the magnetosphere. This causes pressure to build up until bursting point is reached. The material then released crashes into the neutron star to produce the x-ray burst.

The scale of x-ray emission and its rate of occurrence are determined by the critical level required to achieve bursting point. The higher the level the longer it takes for the pressure to build up and the larger the eventual x-ray burst.

HEAO 1

In contrast with the tiny SAS 3, NASA launched in August 1977 the three and a half ton High Energy Astronomical Observatory, one of the largest scientific satellites ever built. It was to be the first of a series of three, thus was known as HEAO 1; and, despite its enormous size was only half as big as its chief scientist, the veteran Herbert Friedman, wanted it to be. Already it was beginning to be evident that budgetary constraints were about to limit x-ray exploration. To add to Friedman's problems, some of the detectors were damaged during flight and HEAO 1 did not live up to its designed specifications.

The object of the flight was to use huge detectors to map new x-ray sources for the next HEAO satellite to examine in more detail. The detectors were built on the same principle as the Uhuru detectors that flew seven years earlier, but by virtue of their size were seven times more sensitive. Despite the damage they suffered, they produced a huge mass of data, much of which is still stored on computer tape at NRL awaiting processing. HEAO 1 did reveal many more sources far out in space, but the most controversial finding it made was not of any individual source. According to the team at Goddard, there appears to be an overall background of radiation in the x-ray band. If this is true, it alters our current conception of the future of the Universe. Put crudely, current astronomical dogma states that the Universe will go on expanding for ever. The diffuse x-ray background which some of the interpretations of the HEAO 1 data suggests, would indicate a diffuse gas spread through the Universe. The mass of this gas could be sufficient to cause the Universe to implode under the influence of its own gravity; this would radically alter our current notions of, so to speak, eternity. These theories, though, and even the interpretation of the data, are disputed between the rival teams with a heat that may almost be sufficient to generate x-rays of itself.

X-ray Telescopes

Even before the first observation of Scorpius X-1, both British and Americans had imagined the possibility of even more direct observations of x-ray sources in the sky: nothing less than images, like the images seen through an optical telescope. But there is one inherent and crippling problem that bedevils such an enterprise: x-rays cannot be focused by ordinary lenses or mirrors because they pass straight through the material. It was thought for many years that they could not be focused at all; but it was just that the impossible takes a little longer.

In the 1930s it had been discovered that x-rays can be reflected if they strike an optically smooth surface at a very shallow angle, so shallow that it almost grazes the surface of the mirror. This grazing reflection could, in theory, be used to focus x-rays, if the mirror could be ground accurately enough to a shape that could focus the rays on to an array of detectors. It should be possible to produce from such an instrument an image of x-ray sources which would reveal detail quite out of the reach of even the best more orthodox Geiger counter type of instrument. But between the theory and the practice was a period of many painful years.

As early as 1959, Robert Boyd and his group had such an instrument in mind. In 1972, thirteen years later, their x-ray telescope flew in the Copernicus satellite. It was a rather modest instrument, focusing rays on to a small detector and building up an image in stages by scanning the x-ray source; but it was the first x-ray telescope to look at the stars.

Across the Atlantic, in Cambridge, Massachusetts, Riccardo Giacconi, then a newcomer to astronomy, had much the same idea as Boyd at much the same time. With his colleagues at American Science and Engineering Inc, he built a grazing-reflection telescope with which the first true x-ray pictures of the Sun were taken in 1963; this attracted sufficient funds for AS&E to be contracted to supply an x-ray telescope for the Skylab mission. With it, the astronauts took sensational pictures of the Sun, and its corona in 1973. The sheer beauty of the pictures was a remarkable achievement; but they also contained volumes of information about the processes in the Sun's corona. Before any pictures of the stars could be taken, though, the old hurdle had to be overcome: how to make an instrument sensitive enough to take pictures of the far, far, fainter stars?

In November 1978, after nearly twenty years of effort, Giacconi's telescope flew up to look at the stars. It was aboard HEAO 2, a satellite now named Einstein. The telescope uses eight mirrors, in four groups of two; they nestle inside each other like Russian dolls. Their surfaces are polished to an accuracy of

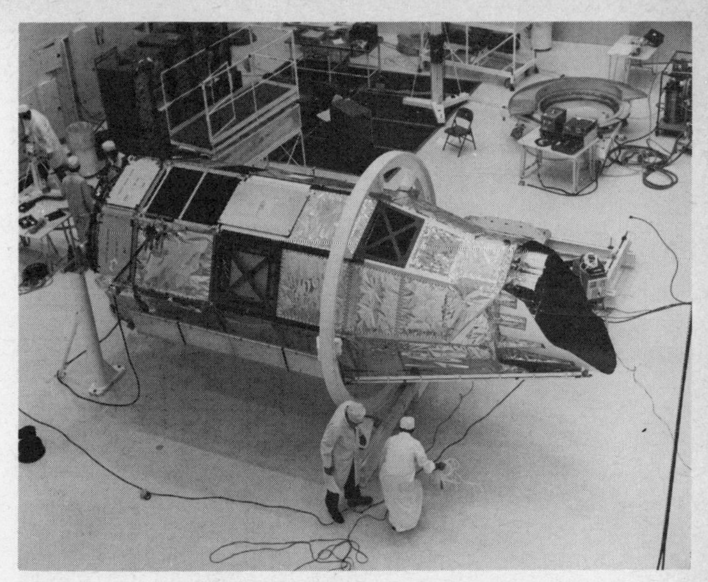

The HEAO – 2 satellite, renamed 'Einstein' after its launch, on test at Cape Kennedy. On the right hand end of the satellite, the brightly polished tubes containing the mirrors of the x-ray telescopes can just be seen.

one-thousandth the wavelength of light and reflect the x-rays to focus on what is essentially a TV camera. The heart of the TV camera is a bundle of thousands of microscopic glass tubes. The inside of each tube is specially coated: when one x-ray photon strikes the coating, it releases an avalanche of electrons which are detected at the bottom of the tube, so that a picture is built up. This part of the instrument is built in the UK under the guidance of the Leicester group. There are two cameras, one with a narrow field of view and high resolution; one with a wider angle of view and therefore lower resolution. The satellite also carries two spectrometers (the kind of instrument that discovered the iron gas in the Perseus Cluster) developed respectively at Goddard and MIT, and the computer software and data analysis is the responsibility of a group at Columbia University in New York.

Einstein

Einstein's Principal Investigator – the boss – is Riccardo

Giacconi. His team is now at the Harvard-Smithsonian Center for Astrophysics in Cambridge, Mass. known as CAO. To Giacconi's delight, but not, one somehow gathers, altogether to his surprise, Einstein has behaved almost perfectly and has shown its paces to the world in the most spectacular way.

Running a big scientific experiment like Einstein is expensive in cash and manpower. As it orbits round Earth, the satellite radios back its data to NASA ground stations at various points. These signals are sent to Goddard, which receives and processes them. One can watch the pictures building up, photon by photon, in the underground control room, where they are recorded on tape which is dispatched every few days to the various experimental laboratories. At the Center for Astrophysics, the tapes are unpacked from their cardboard cartons and checked into the library. Astronomers from all over the world are queuing up to get time on CAO's computers to examine the data. One group in a slightly privileged position, by virtue of their share in the hardware, is Ken Pounds' team at Leicester, who have tapes which they are reprocessing to get rid of electronic noise and produce a 'cleaned-up' picture.

What has Einstein discovered? The problem is: too much. The scale of its operations may be clarified by realising that it is capable of detecting sources ten million times fainter than Scorpius X-1, the very first x-ray star ever observed. With such sensitivity, Einstein can see very weak stars nearby, or stronger sources virtually at the edge of the Universe. Yet there is a paradox in Einstein's superb efficiency. Because its sensitivity is comparable with that of the 200 inch optical telescope on Mount Palomar, pretty much the whole of the known Universe is within its gaze and thousands of sources that were previously undetected and undetectable have been found. Yet so far, no truly astonishing discoveries, like those of the earlier satellites, have been reported. It is as if the first romance of the subject has faded away a little and the explorers now have to settle their new territories and metaphorically clear the forests and plant the crops. There is another frustration: within its planned life, Einstein will simply not have enough time to look at the whole sky. It will plunge to its fiery death with part of its promise unfulfilled.

Of course, Einstein has made many discoveries. With the imaging instruments, the experimenters have been able to see objects at the extremes. Faint stars, nothing exceptional by anyone's standards, turn out to be active emitters of x-rays. One

Two quasars through Einstein's telescope. At the bottom right, quasar 3C 273, first observed nearly twenty years ago. The small bright cluster at top left is a quasar discovered by the Einstein satellite. It is 10 billion light-years away from Earth. Quasars may derive their immense energy from black holes.

class of so-called cool stars turn out to emit a million times more x-rays than was predicted, rather as if an apparent beggar turned out to be Howard Hughes. At the other end of the scale, Einstein's observations may help to shed light on the mystery of the quasers, quasi-stellar objects whose nature is still a puzzle. The most distant quasars emit in x-rays alone as much energy as 1 000 000 000 000 000 suns would emit at all wavelengths, and this within a volume less than a light year in diameter. One possible explanation could be that quasars are powered by giant black holes; eventually these ideas may explain what quasars are, and whether they are the precursors of galaxies, or vice versa. With the help of both optical and radio astronomers, some of these puzzles may in time be resolved.

The Center for Astrophysics believes, too, that it is on the point of disproving the existence of diffuse background x-radiation suggested by the HEAO 1 observers, simply because so many stars are now seen to be x-ray sources that they could contribute most of the observed 'background'. But that is a controversy about which there is no rush to a final judgement. By even the most pessimistic estimates, the end of the Universe is a long way off.

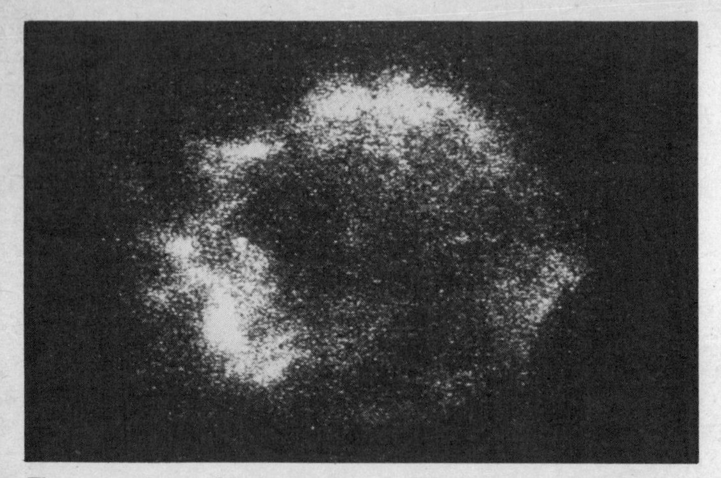

The x-ray emissions of Cassiopeia-A seen by Einstein's telescope. The bright areas show the shell of hot gas emitted after the explosion of a star in the year 1675, a supernova. Unlike the Crab Nebula, there seems to be no central neutron star.

Perhaps the most striking achievement of the Einstein observatory to both laymen and scientists is the beauty of its images. A picture is such a direct way of imparting information; without any knowledge of astronomy, one can instantly grasp the meaning of, say, the glowing cloud of gas surrounding the supernova remnant, Cassiopeia A – the aftermath of an unimaginably violent explosion. The Goddard group's spectrometer, meanwhile, reveals that the material emitting x-rays contains sulphur, silicon and magnesium; visual splendour is translated into analytic information. The MIT spectrometer, too, is probing the chemistry and physics of Einstein's x-ray Universe: the giant elliptical galaxy, M-87 seems to have oxygen flowing in towards its centre. So much information is coming in that the data will take years to process, let alone to understand.

Balance Sheet

In about twenty years, x-ray astronomy has progressed as far as optical astronomy progressed in the three hundred or so years that separate Galileo's telescope from the giant on Mount Palomar. The difficulties have been immense and have been solved triumphantly. In engineering terms alone the Einstein satellite must be one of the major achievements of the twentieth century. The information that it and the other satellites have provided opens a window on processes in a celestial laboratory of such intensity that they could never be reproduced on Earth. They will teach us more and more about the origins of the Universe and so about the origin of the insignificant speck of dust that we inhabit. Every atom in our bodies was made in some star, and we share this inheritance with distant galaxies whose light started on its journey to us billions of years ago.

Twenty years, however short in comparison with astronomical time, is a long time in a human life. As Riccardo Giacconi observes: 'The most difficult barrier, I think for a young scientist to overcome is the realisation that the programme he thought he could accomplish in five years is going to take his entire scientific career. And that's kind of sad, because maybe one could do things a little faster; if realisation came a little bit faster after conception, maybe one could go on to other things. But, as it is, one gets committed for one's lifetime without knowing it.'

But thinking time has been short. In the years since Galileo saw the moons of Jupiter, there has been plenty of time for the optical astronomers to think up theories, get them wrong, go back to look at the stars and think again. The process is still going on. Today's theories may look pretty silly in fifty years.

The x-ray astronomers have not had the luxury of a few centuries of thought. The experiments have been done, the observations have been made, but there is simply too much information to digest in such a short time and new theories, new ideas, are needed to make sense of the data.

A new satellite to suceed Einstein (whose time ran out in 1981) is on the drawing board. With luck, it might fly in 1987 or 1988. But with enormous budget overruns on the space shuttle and shrinking expenditure on new big science, nobody is counting on it. British and European experiments will almost certainly fly first. Perhaps a pause for a few years will do no great harm: a period of consolidation after exploration.

Perhaps there is nothing much new to discover after the avalanche of information that has fallen on the world of astronomy in the last couple of decades; but it would be rather premature to assume that. According to Herbert Friedman: 'Judging by the last twenty years I think one would be ultra conservative to say that the field is finished off, because we've been wrong every time we suggested that possibility. Immediately something incredible has burst on the scene.'

It probably will again.

We have arrived at the outer reaches of the Universe. At least we have arrived at those parts of the Universe furthest from that minuscule planet, Earth. We must never forget that we are travelling backwards in space—time. X-ray astronomy has brought us to the domain of those strange objects known as quasars, blazing pinpoints in the sky which may represent the Universe as it used to be. We shall never know exactly what these things have evolved into by now, but we can reasonably assume that they have reached a similar stage to those parts of the Universe nearer to us and that our part of the Universe used to be like that once. Huge amounts of energetic radiation, all travelling at the speed of light, are confined to relatively small spaces. A quasar can be less than a light year across. But some of that energy has been travelling towards us for thousands of millions of years, so it is hard to conceive of a quasar, or any other object, as having an edge.

But an edge we have indeed reached. As we have travelled back in time, the Universe has in a sense been closing in on us. We are approaching the edge of the primordial explosion from the inside. Recognition of this edge and confirmation of the 'big bang' theory, if any were needed, came through the study not of the high energy end of the spectrum which showed us the quasars, but the cool, low energy, long wavelength end. The slowest frequencies, the faintest, feeblest signals from the sky produced the biggest surprise.

8 A Whisper from Space

Peter Jones

The Space Between the Stars

Looking at the starry sky, it is a commonplace to say that the space between the bright stars is dark. If we look beyond the stars of our galaxy to the distant galaxies, the space between them is dark too. And yet the most important discovery of the last fifty years in our study of the Universe has been the remarkable fact (first noticed in 1965 and since confirmed) that in what appears to the eye as the blackness of space, there is invisible radiation. The whole sky is glowing brightly. How this was established and why this provides us with such an extraordinary insight into the Universe, is one of the most intriguing tales of twentieth-century astronomy.

Throughout earlier centuries, the eye and the mind behind it led us to an accurate but limited understanding of the Universe in which we live. In the last few decades, however, a new Universe has been revealed to us. The invisible radiation hidden from our eyes has become clear through the cunning instruments which sense radio waves, x-rays, ultraviolet, infrared and microwaves. For example, the great family of big dishes which focus invisible radio waves is familiar: they can see what the eye cannot. We can speak of seeing because all the waves we shall discuss are close kin to visible light. They are all part of the electromagnetic spectrum and differ from the light and colours we can see only in wavelength, as indeed red light differs from blue or, by analogy, a tone of low pitch differs from one that is high only in the wavelength of the sound.

The range of vision of the human eye is analogous to a piano keyboard on which just a few notes in the middle can be played. Ultraviolet, x-rays and gamma rays are in the upper registers, infrared in the lower. Furthermore, six feet beyond the bottom end of this imaginary piano, there is another range of invisible unheard keys – the scale of the radio astronomer. It is in a small

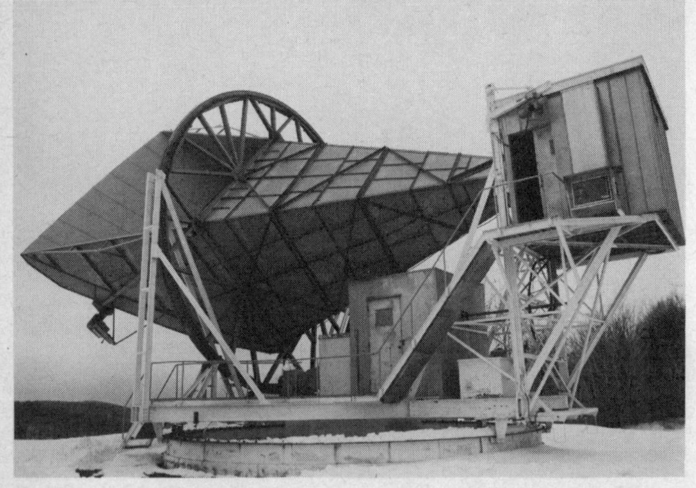

The Horn antenna at Bell Telephone Laboratories, Holmdel, New Jersey. Among all the listening 'ears' in the world it was this one that caught the crucial whisper left over from our eruptive creation. The horn feeds its signal to a sensitive amplifier. This, kept very cold to reduce internal disturbance, is in the little box-like structure to the right.

part of this scale, maybe one octave, that microwave radiation has been detected: this has transformed our knowledge of the cosmos.

The Holmdel Horn

The detection of cosmic microwave radiation required not only a uniquely specialised instrument, but operators of rare sensitivity to both machine and sky. Both men and machine were already together in 1964: they were to be found at the Bell Laboratories in Holmdel, New Jersey. There, a microwave horn looking like an old-fashioned ear trumpet for a hard-of-hearing giant, still sits upon a hilltop. A most unusual antenna, it had been built for communication via the Echo satellite and had been designed from the outset to catch microwave signals, radio waves as short as the width of a hand. The horn funnelled the microwaves to a sensitive amplifier, kept very cold to reduce internal disturbances.

A pair of young radio astronomers, Arno Penzias and Robert Wilson, joined Bell Laboratories with the promise that, if they applied their talents to perfecting the antenna for communications, they would later be allowed to use it for astronomy. In 1964, they embarked on an overall survey of the radio noise coming from our own galaxy, in effect from the sky itself. They chose first a seven centimetre wavelength because they expected little signal there. It was a waveband people thought free of most galactic emissions and should have provided a quiet baseline. They expected almost no radiation from any part of the sky. 'Instead what happened is that we found radiation coming into our antenna from all directions,' Wilson said. 'It was just flooding in at us and clearly was of an order or magnitude more than we expected from the galaxy.'

This was, to put it baldly, an embarrassment. Penzias and Wilson had a result which they knew could not be right. They believed that they could confidently discount any radio source known to exist, for all radio sources known at that time radiated at longer wavelengths rather than at shorter. Yet here was radiation flooding in at a relatively short wavelength. Anyway it was coming from everywhere and there was much too much of it. Was it a false reading? The compact amplifiers were easily tested: maybe the big horn antenna was at fault.

Suspicion first rested upon the throat of the antenna. Its point of entry into the cab was always slightly warmer than the surrounding air, and particularly in winter it became a favourite roost for pigeons. Although the birds flew away at the arrival of the astronomers, they had coated the throat with their droppings and both Penzias and Wilson feared that these droppings might not only absorb but also emit all or part of the radio waves that they had received.

However, the pigeons had to be acquitted when, after dismantling, cleaning and reassembling the antenna, most of the effect, surprisingly, remained. Suspicions now turned to the horn itself. Could the sheet metal structure be emitting some unwanted radio noise? Painstakingly the experimenters then resealed every joint in the big structure with metal tape to be doubly sure. The signal remained unceasing. Almost reluctantly they had to recognise that the signal was coming somewhere from outside the horn. But what was its source?

All that remained was the sky itself and, however hard it was for Penzias and Wilson to accept it, there was no alternative. They had therefore begun by checking the Sun and searching for something in the solar system; but the signal remained constant without seasonal or nocturnal variation. Nor was there any

Bob Wilson and Arno Penzias (foreground) *search the horn antenna for structural clues to the mysterious noise. To ensure that the sheet metal structure was not emitting some unwanted noise, the experimenters painstakingly resealed every joint in the big horn with metal tape.*

change, no matter in which direction they scanned the sky. Having eliminated the possibility that this radiation might be polarised, the next larger object to be considered, the only remaining thing, was our own galaxy.

The galaxy itself is somewhat saucer shaped and we are within it. If Penzias and Wilson looked towards the centre of this saucer they would expect a bigger effect than if they looked away or if they looked perpendicularly to it. Yet no such change was evident; nor did any galaxy similar to our own emit any radiation of this magnitude. The only conclusion that they could draw, astonishing as this might be, was that this radiation was coming from somewhere in really deep cosmic space. This was something that neither had known or even dreamed of, since they were unaware that there were a few people who had such dreams and had even conceived of such radiation twenty years earlier.

Cosmological Dreams

In the late 1940s, George Gamow had begun a theoretical exploration of the origins of the matter of the Universe. To do so he had to conceive of the very creation of the Universe, a process in the course of which, he believed, atoms of the Universe were forged. To his mind, the Universe began in a gigantic explosion – the 'big bang' theory was born. It seemed incidental to this great theory at the time that, if the Universe was once very hot and had been cooling ever since, then the energy of the original 'flash' ought still to have been around, though greatly dissipated. Nevertheless in 1948, Ralph Alpher and Robert Herman, two colleagues of Gamow, explicitly predicted the existence of relict primordial radiation coming from all directions at almost equal strength. As Gamow himself dramatically expressed this: 'We live in an ocean of whispers left over from our eruptive creation.' Sadly, no scientists considered searching for this whisper: few scientists at that time knew how to listen. Within a few years Gamow's big bang team disbanded and went into other areas of science.

In so doing, this became one of the 'missed opportunities' of science, for the technology for the search did indeed exist. In the same volume of *Physical Review*, the journal in which Gamow's 'big bang' theory was published, one may find within a few pages, an extraordinarily pertinent report from the redoubtable innovator and experimenter Robert Dicke. Building on wartime studies which had led to the development of the Dicke radiometer, an instrument for detecting microwaves, Dicke and several colleagues published in 1946 a study of microwave radiation emitted by atmospheric water, which also provided as a by-product an upper limit on cosmic radiation. But in reporting on microwave cosmic radiation, Dicke had no link in mind with the 'big bang' theory, and as a result, his group and Gamow's never made contact.

In the same way, cases of independent investigators missing a vital piece of the jigsaw continued to characterise our story. In 1964, two Soviet astronomers named Doroshkevich and Novikov realised that Gamow's 'ocean of whispers' might be observable. They even suggested that, of the world's radio antennae available for the job, the best suited was one operated by Bell Laboratories in Holmdel, New Jersey. At that time, and unknown to either the Soviet scientists or the Bell scientists, Robert Dicke had returned to the search for cosmic microwave radiation. He now worked with colleagues at Princeton University, just thirty miles west of Holmdel, where Penzias and Wilson continued to be baffled by the unwanted microwaves of the horn.

Rival Theories

Dicke is renowned for insight and skill in settling long theoretical arguments by sensitive and novel experiments. In 1965 he had just persuaded his colleagues that a real clue about the early Universe was within experimental grasp. At this time of open cosmological speculation and too few observations to settle disagreements, there were many alternative theories of the origins of the Universe. As Jim Peebles, Dicke's colleague at Princeton, has put it, 'You start with a fundamental idea, and branch into possible alternatives. But each alternative branches, in turn, into possible alternatives. Because the Universe is a complicated place, any reasonably inventive scientist can think of all sorts of possible Universes, and you can end up with a bewildering variety of Universes to think about. It's long plagued the subject, which is one reason why there are such violent debates in it.'

Centre stage of the sometimes almost theatrical debate were two main contending theories: the steady state and the big bang. Supporters of each theory accepted the premise of galaxies that are flying apart, but formulated radically different descriptions of the expanding Universe.

Three British astronomers, Hermann Bondi, Thomas Gold and Fred Hoyle, had put forward the steady state theory in 1948. They doubted whether, in a 'big bang' Universe, the laws of physics could remain unchanged: the possibility that they might not seemed an unacceptable proposition to Bondi, Gold and Hoyle. Their alternative was a Universe which would appear never to change with time. They proposed that new material is steadily, or continuously, created to make up for the progressive thinning out of the material in the Universe, thereby producing a Universe that neither changes with time, nor according to where it is viewed.

To these theories Dicke added a variation on the 'big bang'. There were two possible views to take of this theory, which was that all the matter that was together in the beginning of the Universe was simply created; that was the start, and there was nothing before it. There was another view which is that the Universe that we see expanding is a remnant of some earlier Universe that had collapsed, and then expanded again.

An Echo of the origin of the Universe

Whether you had an oscillating Universe or a simple big bang, Robert Dicke and the brilliant theoretician Jim Peebles soon saw the importance of one particular feature of the early Universe. The fire ball would be so hot that it would endow the Universe with a large amount of radiation to start with. It was Dicke's hunch that the radiation would still be around today and could be searched for. The Princeton group might well have turned now to Gamow's calculations, but ignorant of what had already been done, they started from scratch. Jim Peebles, who had the task of working out the details, was thus faced by this critical question: if this radiation was present how would they be able to detect it and how would they know that they were detecting it and not radiation from some other source in the Universe? After all, our own atmosphere, own own galaxy and other galaxies are all among the many radio sources emitting radiation. They had no guarantee that this radiation might not confuse the picture, so that even if they did detect radiation from the 'big bang', they might never be able to convince themselves that they really had detected *it* and not radiation from something relatively local.

This critical question was to underlie much of the research of the following years, but Peebles' early calculations showed there was a chance to pick up the signals in a waveband free of interference. So, in the summer of 1965, on the roof of the Physics Lab at Princeton the experimenters, Roll and Wilkinson, got ready their microwave horn. They stretched a net for shorter wavelengths than the fish caught at Holmdel.

In fact at Holmdel, only an hour's drive from Princeton, the experiment had already been done, unconsciously, but very well. The two groups knew nothing of each other, but that was about to change. Peebles had been invited to talk at another university about the planned Princeton experiment. His colleagues did not object. As Peebles explains, 'They gave me leave to talk about this experiment and about our ideas because it was clear that nobody could catch up with us. What we didn't anticipate, of course, was that someone might have already done the experiment, and not fully appreciated the answer.'

The go-between who attended the lecture was MIT radio astronomer Professor Bernard Burke. 'I was just the marriage broker. Peebles had just spilled the beans about this marvellous experiment that they were sure could only be done at Princeton. When I then got a telephone call at lunch, from Arno Penzias on

In the summer of 1965, on the roof of the Physics lab at Princeton, the experimenters Roll and Wilkinson got ready their smaller horn (top centre). *At the time they didn't know that just 30 miles away at Holmdel, the experiment had already been done, unconsciously, but very well.*

an entirely different matter, it crossed my mind that Arno was doing at Holmdel just the measurement that Peebles had been talking about. So I said, "Arno, how's that crazy experiment coming along?" And he said, "Well we have some results that we don't quite understand." And so I said, "Well why don't you call Bob Dicke at Princeton and I think he can tell you."'

Jim Peebles can remember that call. He heard the discussion in the background, bits and pieces of it, and could not imagine what was happening. Then Bob Dicke came back and said casually, "Well, boys, I think we might have it."'

The news was out. The Holmdel whisper was no less than an echo of the primordial explosion – the Big Bang. On 21 May 1965 the discovery made the front page of *The New York Times*. Science correspondent Walter Sullivan's story began: 'Scientists at the Bell Telephone Laboratory have observed what a group at Princeton University believes may be remnants of an explosion that gave birth to the Universe.' This account, Robert Wilson later declared, was what really brought home to him the full magnitude of the discovery. Until then, he and Arno Penzias had simply not absorbed its importance.

Meanwhile, in *Astrophysical Journal*, papers from both the Bell and Princeton teams were published jointly; the former reported their discovery and the latter offered an explanation. Thirteen years later however, it was only Penzias and Wilson who received the ultimate accolade of a Nobel Prize.

The Blacksmith's Shop

How can we be sure what that whisper from space is saying? We can begin by considering the simple physics of the blacksmith's shop, for the radiation that Holmdel sees is known to be heat radiation, the very same emissions that come also from the fiery forge heating the blacksmith's face.

This physics, turn of the century physics, is a foundation stone of the modern theory of matter and energy. It says, in short, that everything that has a temperature radiates heat. A blacksmith makes up a chain link by link. When the blacksmith handles the newest red hot link, his eye warns him to use tongs. The luminous link evidently radiates; but even the links made a while ago, and already black, are also vigorously glowing, but with a radiation invisible to our Sun-adapted eyes. Indeed, it is possible to measure such radiation using a wide band detector which is not at all like the eye. Everything in the Universe at any temperature radiates, even at the temperature of liquid helium 3 degrees above absolute zero, that is, 270 degrees below the freezing point of ice. Only at absolute zero do things not glow at all to any detector.

The Signature of Heat

We are so partial to sunlight and its manmade imitations that it is hard to accept that objects radiate all the time. So it is not surprising that our understanding of heat radiation has been gained only with the ingenious detectors which extend all our senses. Using these it is possible to show that, hidden in the radiation from an object, is a clue to its temperature which is as unique as a signature. A simple experiment can demonstrate this.

We take the coiled filament of an ordinary light bulb glowing with white heat, the heat provided by an electric current. We pass the light through a prism which draws out the component colours into a rainbow stripe. We then pass a suitable wide band detector, one less biased than our eyes, across the rainbow and beyond, to record the radiation intensity colour by colour. The impartial representation is remarkable in several ways. The detector records radiation long after it has passed the last we see in the red, and the peak of the radiation, the maximum given out by the filament is in the infrared, the radiation known to physicists for 150 years.

This commonplace hump is in fact the signature of the glowing light. The peak occurs uniquely at a particular place in the spectrum for a particular temperature. This we can show by varying the electrical power of the lamp and in so doing varying its temperature as well. It is then readily seen that each particular temperature has its own individual curve, its own distinctive peak.

When the potter loads the cold, dark kiln, it is hard to believe that it too has its characteristic signature: but it does. Light up, and very slowly the kiln comes to red heat and beyond. The radiation becomes a glow visible to our eyes. At every stage its temperature is characterised by a unique hump. Watch closely and an unexpected property of heat radiation appears. As the load in the kiln reaches steady, even temperature, it is not just the pots that glow, the glow pervades the whole space. All detail is lost. The early Universe itself was just such a uniform and glowing kiln, revealing its temperature in its hidden signature.

A Camel's Rope

All sources of radiation in the Universe peak at frequencies which depend on the temperature of the source. For example, the radiation from the Sun whose temperature is 58 000°K is peaked in 'visible' wavelengths. An intensively hot object of 1 000 000°K would peak in the ultraviolet. The relationship underpinning this comes from the work in the late nineteenth century of the German scientist Max Planck: it is a universal formula, applies irrespective of the nature of the source and is dependent only on its temperature.

The Holmdel experiment listening on just one wavelength, gave us just one point on a temperature curve. In a few months Princeton had another at 3.2 centimetres wavelength. Then the radio astronomers of Cambridge brought in a third at 20.7 cm. All three points fitted the same curve, so that one could read off the temperature of the radiation source. It was very cold, only 3 degrees above absolute zero. All these points were far below the peak of the curve they seemed to fit. The curve was still only presumed. Could the judgment be premature? The sceptics could rightly say this was dangerous because many curves might look like the real thing. Professor Philip Morrison of MIT expressed the critic's point of view:

> We were in the position of somebody sleeping in a tent in the desert, and noticing a camel rope, beautifully braided, lying under the edge of the tent. Well there was no question that it was a camel rope all right. But was there a camel on the other end of it? My view at the time was that you can't say you've caught a camel until you can see the hump. And so the game for me was that we had to follow the curve over the hump of the signature, and not be content with our three points on the edge, even if they fitted mathematically very nicely.

Over the Hump

For the experimenters it was a heavy burden. The millimetre waves of the hump just cannot reach the ground through our watery atmosphere. They would now have to rule out ground-based observations and go outside the atmosphere. One brief rocket flight gave a result far above the predicted hump and delighted the sceptics. The rocket turned out to have looked at the radiation from some of its own heated parts.

The stratosphere balloon was to be the better choice. It soared high enough to leave most of the air below it and it stayed up overnight, long enough for some careful checking.

Many launches were made by a series of labs and slowly the results began to fit the hump. Professor Rainer Weiss's lab at MIT first caught the real peak. Oddly enough, it was realised shortly after the Holmdel discovery that optical astronomers had long had evidence of the cosmic microwave or background radiation, as it came to be known. Of the various molecules that have been detected in space, cyanogen, a compound of one carbon and one nitrogen atom, exhibited an intriguing property that revealed what was an unexpected warming effect in space. Cyanogen adds faint absorption lines to the spectra of stars whose light passes through the gas before reaching Earth. Two of these absorption lines are only produced, however, when cyanogen molecules are bathed, or warmed as it were, by radiation at 2.6 millimetres. This happens to be a wavelength that lies close to the hump of the background radiation, but until 1965 nobody had any reason to associate this mysterious 'warming of cyanogen' with radiation that stems from the origin of the

Universe. For those who were making direct measurements, each experiment meant a couple of years' hard work, usually based on some new detecting scheme or some new device for improved reliability at a new waveband.

In 1977 it was a University of California group at Berkeley who brought back the first evidence which confirmed that once over the hump the radiation energy density then continues to diminish. Point after point, the experimental data moved up and over the hump and down again. It could not be denied any longer: the curve fitted beautifully the signature of 3 degrees K temperature radiation. Moreover, it was absolutely uniform in all directions. Like the radiation inside the potter's kiln, detail was washed out. It was a remarkable result, but more than this, it was possible to reckon the amount of energy there was in the hump.

In the bright stars, the quasars, the violent events, the supernovae, we have brilliant, hot and extraordinarily rapid emitters of radiation. But the space in between, through which flows the background radiation, is very large. It turns out that the quiet microwave whisper is 99 per cent of all the radiation there is. From the point of view of the Universe there is only the 3 degrees K background radiation: what comes from the galaxies and stars is relatively unimportant.

This is such a striking result that we must give it a fundamental interpretation. We are actually looking at the origin of the Universe. But we must continue to pose the questions that we asked earlier. How can we be sure? How do we know? How can all this energy have come from a source which we do not see; which is uniformly everywhere and yet which is giving us radiation from a source at the very low temperature of only 3 degrees K above absolute zero?

The Red Shift

It was the Austrian physicist Christian Doppler who first pointed out 150 years ago that a change of pitch could be expected whenever a steady source of waves moved with respect to an observer. It is a common enough experience these days – from the roadside a fast approaching car sounds higher pitched that when it is going away – and this is the rule that we now know as the Doppler Shift: approaching, higher pitch, shorter waves; receding, lower pitch, longer waves. The effect for sound waves was first tried by a Dutch physicist in the flat lands of Holland near Utrecht, shortly after Doppler published. Trumpeters blowing a note from an open railway wagon were drawn by a steam engine up to and past observers with a musical ear, standing beside the track. They listened for, and heard, the predicted change of pitch.

It comes as no surprise – Doppler himself expected it – that the principle is the same for light and radio as it is for sound waves. During the nineteenth century, astronomers found it in stars, some approaching, some receding. As a star moves away, its wavelengths becomes stretched, attenuated. So its natural colours seem shifted more to the red. It was this technique which added dimension and motion to our map of the Universe.

About the time of the First World War, first Vesto Slipher, and then Edwin Hubble, found something remarkable. They looked at galaxies, not individual stars, and in looking at these great collections or archipelagos of stars, they found enormous Doppler shifts, always receding, always red shifts.

A Journey to the Origins of the Universe

Quite noticeably, and with great regularity, the galaxies were seen to behave according to a startling law. The farther away, the greater the red shift. This regularity – the farther the distance, the greater the speed – is the Hubble Law of Red Shift, the foundation of modern cosmology. Hubble had created a picture of the Universe as an expanding swarm of galaxies and recent work has largely supported this view. Nowadays the remoter galaxies have even greater red shifts and the quasars are thought to be even farther, receding even faster. As we look at these remote objects we are also looking ever more deeply into the history of the Universe. Building on this we can now offer a firm interpretation of the whisper of 3 degrees radiation: if we are right, it is radiation that is red shifted by a factor of one thousand. It is the record of an invisible glow of most ancient heat, deep in the Universe, from the remote past and farther away than anything else we can see in the Universe.

Let us move in our imagination at breakneck speed and follow the radiation back to its source: a journey that takes us back in time as well as in space. We soon leave the solar system behind. Beyond it, we reach the bright star cluster Hyades. Far beyond it is the edge of the Milky Way. When we leave this 'our galaxy', we enter the realm of other galaxies, great archipelagos of stars. The radiation has travelled for aeons to reach us, so as we move outwards so we go back in time. We enter a region where as yet we cannot see anything at all, which is at present beyond the telescope. Farther still, and we are at the very time when the galaxies were being formed – it is a blank space and unknown.

Still earlier and farther, we reach the misty source of the Holmdel whisper. All we can see is a white hot wall of hydrogen gas. It is hot, not at a few degrees, but at 3000 degrees or more. We see it from Earth intensely red shifted and therefore cooled a thousandfold. We have reached the early Universe, when it was perhaps less than a million years old.

It would in some sense be like living inside a sunlit cloud, bright everywhere, but not brighter on one place than another. And impossible to see very far, not because it was not light, but because the light is scattered randomly from one atom to another, throughout the gas. We can think of that early Universe as a vast, glowing neon tube, filled not with familiar red neon, but the simpler primordial hydrogen. In such a glowing gas, most of the atoms are broken into their constituent electrons and protons.

As the early Universe expanded, that bland 'plasma' as it is called, slowly grew cooler and cooler; 4000 degrees; 3500 degrees; 3000 degrees. At that point a remarkable transformation occurred. When the temperature was no longer high enough to break apart the atoms of hydrogen, each electron, finding a proton, joined in a tight dance with a proton as a hydrogen atom. Once that happened, the electron mist was no longer opaque: hydrogen is transparent. Radiation was then set free from matter and has been glowing separate from matter ever since that time.

A Relic of the Most Ancient Past

Nature has been generous in showing us the 3 degree radiation. Think how remarkable it is: it comes from a single time, the time when the separate electrons and protons of the early Universe combined to make transparent hydrogen; it is marked with a single temperature; it comes uniformly to us from all directions and it must be the same throughout the Universe.

As the radiation travels through space in an expanding Universe, it is successively reddened by the red shift to longer and longer wavelengths. These longer wavelengths shift the hump in the characteristic signature to lower and lower energies, so the measured temperature becomes smaller and smaller gradually with time.

Had we measured a hundred million years ago, we would have

found a little more than 3 degrees. Were our descendants to measure a hundred million years from now they will find a little less than 3 degrees. The background radiation provides in this way a kind of cosmic clock which every observer on every galaxy, if there are observers amongst the distant galaxies, could calibrate one with the other the Universal time since the 'flash' of radiation was released from the plasma of the early Universe.

That moment was between fifteen and twenty thousand million years ago. (The Universe is about four times as old as Earth.) Since that time, matter derived from the hydrogen has condensed into the complex Universe of lumpy galaxies in which we now live. So now the black space is full of radiation and the shiny points mark the condensation of matter.

Only 5000 million years ago, the background radiation bathed an insignificant new star, one among many. That was our forming Sun, surrounded by planets. Today, we are still bathed in the radiation as it continues to grow steadily redder, cooling as time passes. It is the relic of the ancient past, a fossil more spectacular than any dinosaur bone. It deserves study for every little detail, every tiny feature we can find, for it may speak to us of a time we cannot otherwise know. It might tell us for example of the time when matter and the background radiation went their separate ways. It might tell us where and when the matter condensed, for small differences in the otherwise uniform radiation might be the sign of where this 'lumpy' Universe formed. And it might tell us, too, how our part of the Universe is moving relative to the background radiation, for we can invoke the Doppler effect to study the moving observer as well as the moving source.

In 1977 and 1978 a search for such fine detail was carried out by a group in the Radiation Laboratory at Berkeley. They flew high in the air, not by balloon, but in an old U2 spy plane adapted to look up into the Universe rather than to spy down into a hostile territory. The U2 was fitted with a pair of small open receiving horns matched to millimetre waves. Two horns enabled the investigators to compare one direction with another to see if the radiation showed any sign of directionality. The horns rotated to exchange places and cancel out any inbuilt bias.

On the first few flights it became apparent that the radiation was not quite uniform in every direction. With more flying, a pattern made itself evident. The radiation was a little less red shifted in the direction of the constellation of Leo and, significantly, it was a little more red shifted 180 degrees away in the direction of the constellation of Aquarius. Between these regions the variation was smooth. This offered the ready interpretation that the variation was not an intrinsic variation in the background radiation itself, but was due to the motion of Earth through the background radiation. We are in effect disclosed by the Doppler shift as moving observers. Our speed in the Universe is about one part in a thousand of that of light – about 250 miles a second. Once we allow for the Sun's great orbit as it swings around the Milky Way, we can conclude that the centre of our Milky Way is itself moving at a modest celestial speed. In roadside terms, it is a million miles an hour or so, which somehow our whole galaxy picked up during formation.

Myths – Ancient and Modern

The background radiation has given us a great deal, not least a magnificent benchmark for determining motion in the Universe. But remember the kiln and how, as it glowed more and more strongly, the detail was washed out. In the same way, the early Universe presents to us a veil, an opaque veil of glowing plasma, which is what the background radiation seems to represent. It may be that we shall never see farther in to the opaque plasma than we do now. This does not mean that it is a wall impervious to knowledge. We have other means, other radiations, neutrons, gravitation – who knows what will come.

But the idea that out of a bland homogeneous background all the complexity of the world could arrive is not an idea entirely new to astronomy. It is an idea as old as the Sanskrit philosophers. Long ago, in part of the creation myth, they told how Vishnu the Preserver, with a mighty churn made of a mountain, churned the Ocean of Milk, and out of this bland and nourishing material came the Divine Cow, first nourisher of mankind; an eight-headed steed; also the Moon itself, and many other such wonders.

To Professor Philip Morrison there is a logical similarity between the myth makers and ourselves who make a grand myth out of the substance of science. Both are inquiring how one might make the complex from the simple. Those persons long ago saw their metaphor in the act of the kitchen where every morning with a simple churn, the butter could be made out of the uniform milk. We look on a much grander scale, at a much more impersonal and distant sea – a sea of hydrogen plasma.

But in some way, by forces such as gravitation, rotation, tidal effects, which we are only just beginning to understand in some

Churning of the Ocean of Milk: in part of the Creation myth, Sanskrit philosophers told how Vishnu the Preserver, with a mighty churn made of a mountain, churned the Ocean of Milk. Out of this bland material came the divine cow, first nourisher of mankind, an eight headed steed, the moon itself, and many another such wonders.

detail, there was formed out of the bland, uniform hydrogen, the lumpy, complex mix of galaxies and the stars among which we observed and of which we are part. Through the discovery of the background radiation, we have a secure platform from which to explore and, standing on that platform, the cosmologists of the future will be able to go into the deeper past and discover more about the origins and directions of the Universe at large.

Having reached the edge of the Universe we can now look back. Standing in front of the opaque curtain which conceals the 'big bang' itself we are, perversely, looking forward in time. We can see the next seventeen or eighteen thousand million years spread out in front of us. From here it is plain that there is one common object which is the building block of the Universe – the galaxy. It is within galaxies that we find the neutron stars, the pulsars, the x-ray sources and the invisible black holes that may lie at the centre of some galaxies. In some parts of galaxies new stars are being formed; elsewhere old ones are either exploding in supernovae or collapsing into white dwarfs. There are binary star systems and single star systems, one of which at least is encircled by a system of planets, one of which supports life.

This has been an impossible journey but it has, hopefully, revealed much. It is worth therefore returning instantaneously to base so that we can look out once again at this astonishing Universe with the benefit of hindsight. We cannot look at our own galaxy with any ease because of our position inside it, but by observing from Earth the truly amazing array of galaxies, we can choose a mirror in which to look. At the same time we can pay tribute to some of the astronomers and physicists who have peopled these pages and have enabled this journey to be undertaken. We may not be able to foretell our destinies but, thanks to them and to their discoveries about the galaxies around us, we can know and understand our origins, if not in sparkling detail, at least with a degree of clarity unimagined just twenty-five years ago and unimaginable a hundred years ago.

9 Beyond The Milky Way

Alec Nisbett

To look out into space is to look back in time, in to the history of our Universe. Travelling at 186 000 miles per second, light takes several years to reach us from the nearest stars; thousands of years to travel from vast, redly growing clouds of hydrogen in which new stars are forming; and 30 000 from the brilliant congregation of yellow stars shot with dark lanes of dust that marks and obscures the centre of our own Milky Way Galaxy.

Our galaxy contains perhaps a hundred thousand million stars. Astronomers have plotted its shape. Imagine a ping-pong ball set at the centre of a disc some twenty inches across and half an inch thick. The predominantly yellow nucleus is composed mainly of ancient stars that have been there since the galaxy first formed. The surrounding disc contains a spiral frosting of stars. Many of them are huge, young, and brilliantly blue, and some like our own Sun are much smaller, in undistinguished middle age. Fewer in these outer regions are older, first generation stars like those at the hub. The regular spiral of the Milky Way Galaxy is a hundred thousand light years across its visible diameter and has two irregular companions. These are the Large Magellanic Cloud, which has a quarter of the number of stars in the Milky Way and its spirals, and the Small Magellanic Cloud, which has a sixth.

Travel out from our solar system through the nearby spread of stars that forms the Andromeda constellation; continue outward beyond our galaxy to a distance over twenty times its diameter . . . and another will at last loom up – the Great Galaxy in Andromeda. A spiral even greater than our own, it also has two smaller companions, fuzzy elliptical concentrations of starlight that contrast in shape with the huge, flattened disc that echoes the Milky Way. The Andromeda Galaxy can be seen from our northern latitudes hanging low in the sky on clear, dark autumn evenings, a faint, misty smudge that for the naked eye must represent the outer marker of the visible Universe. There is one

A spectacular spiral that we can never see – the Milky Way Galaxy modelled by Harvard historian of astronomy Owen Gingerich, who indicates our own undistinguished location in one of its outer suburbs.

other spiral that lies a little further from us, in Triangulum, a sector of the sky not far from Andromeda, that some have also reported seeing by their unaided eye. The light from these two began its journey two million years ago, before man walked on Earth. But that, even for our own planet, is a very recent event, and our local group of galaxies a mere handful of hamlets in the Universe at large.

Beyond our neighbour in Andromeda lie as many other galaxies as there are stars in our own. There are dozens of galaxies for every human being in our overpopulated Earth. In their catherine wheel or rugby football regularities of form or erratic salt-and-pepper sprinklings, they hang at every angle in space. Each is as distinctively individual as a human face. They lie scattered unevenly in the remote dark of the night.

What order undelies this apparently random distribution? How were the galaxies formed; how have they evolved; and what is their future? The answers to these and many other questions are still far from clear, but a shadowed picture is beginning to emerge. This is the story of how some of the key discoveries were made and of the trend of current investigations. Each new

Our most spectacular neighbour, the Great Galaxy in Andromeda, and its two smaller elliptical companions, extend over much more of the sky than the moon, but are only faintly visible to the naked eye. Their distance from us was the subject of the Great Debate – in truth a rancorous dispute – over the nature and scale of the Universe.

advance is linked to some development in technology. In extragalactic astronomy, the fertile power of our human imaginations has been repeatedly outreached by the extraordinary observations that have emerged as each new instrument has probed farther across the Universe and deeper into its history.

The Leviathan of Parsonstown

Our story begins in the gardens of a stately home in central Ireland. At Birr Castle in 1845 William Parsons, the third Earl of Rosse, completed the construction of the largest telescope in the world. Two high, parallel walls supported a wooden tube some six feet in diameter. One end pivoted in an iron socket, the other could be swung freely within the limits imposed by the walls on either side. In order to point the tube continuously at some object in the sky as Earth rotated beneath it, three men would haul it by block and tackle as directed by the astronomer of the night.

Rosse had chosen to build a reflecting telescope, adopting a configuration that Newton had developed from a design by his

Scottish contemporary, James Gregory. Light was collected by a primary mirror at the bottom of the tube and returned to a much smaller secondary mirror, set at an angle near the top. There the light was reflected out to an observer standing on the long arc of a wooden walkway that could be shifted bodily from side to side with the telescope tube. With sketching paper in hand, Rosse peered into the eyepiece.

As the clouds parted briefly about his new instrument on the first night of use (Ireland is not the ideal place in which to build a telescope) he glimpsed the object he had chosen to observe and made a preliminary sketch of its structure. What he drew, for that first time ever, was a spiral. 'The Whirlpool', as it has been nicknamed, seems to have intrigued Rosse: after delays caused at first by the weather and then by his fight against the Irish Potato Famine, he drew the Whirlpool again and again. There are, in fact, two galaxies in its system. The larger is seen flat on, with distinct spiral lanes, on which one reaches out to the bright, compact smudge of the companion.

Rosse observed and sketched many nebulae, as all extended objects outside the solar system were called, and was the first to demonstrate that the spirals formed a special class, though he had no idea what they might be. Indeed, some of his sketches show lenticular shapes that to a present-day visitor to Birr are unmistakably spiral galaxies, seen edge on. The handwritten legends attached to them contain no suggestion of their likely structure in the third, unseen dimension.

Islands of Starlight

Rosse's recognition of the spiral structure was one possible starting point, but not the earliest. The German philosopher Immanuel Kant had already suggested the concept of island universes or, as we would now describe them, galaxies. However, this happy idea was based on an inaccurate report of astronomical observations, so it should perhaps be classified as a lucky guess.

Friedrich Wilhelm Herschel had been smuggled out of his native Hanover to avoid military service and in England became one of the greatest observers of all time and later 'Sir William, the King's Astronomer'. By estimating the densities of stars in different directions, he made a valiant attempt to determine our own Sun's position within a flattened disc of stars spreading from the Milky Way. Gazing into his 49 inch reflector, he also

The 'Whirlpool' (M51 and its companion M52) was repeatedly sketched by the Earl of Rosse, who pioneered the study of spiral nebulae, distinguishing them from the amorphous gas-clouds of our own galaxy.

M51 in a time-exposure on the Kitt Peak/Cerro Tololo telescope. Conventional photography lacks the capacity of the human eye to distinguish between tones and interpret structure.

supposed for a while that, beyond the stars, he was seeing vast numbers of more distant universes (other galaxies). Later, he went off that idea. But his published catalogue giving the positions of many nebulae provided a valuable checklist for Rosse to sort into amorphous clouds and spirals.

Another step forward was made by an amateur chemist and astronomer William (later Sir William) Huggins of Tulse Hill in south London. The lens at the top of his refractor telescope pointed out through a dome on his roof; at the eyepiece he placed a spectroscope. Long before, Newton had shown that the Sun's light, passed through a prism, was split into a rainbow range of colours (today, the hot filament of an electric bulb produces light that can also be analysed in just this way). Chemists, in the laboratory, had found that particular, sharply defined spectral lines – narrow gaps in the continuously changing coloured strip – were characteristic of individual chemical substances. These were called absorption lines. In some other spectra, the continuous rainbow spread was replaced by glowing coloured lines known as emission lines (today we see such lines in the spectrum from a gas-discharge strip-light). Huggins' spectroscope showed that stars, like our own Sun, had dark absorption lines in their spectra. It became possible to identify some of the materials in the cooler outer regions through which their continuous-spectrum, incandescent light had passed. Amorphous nebulae, however, had the bright emission lines: the signature of glowing gas clouds.

Eventually, Huggins turned his attention to the spirals and found that their light was distinctively different from most other nebulae; the spectra of the spirals were more like that of starlight. This strongly suggested that they were actually made up of stars. But no individual dots could be resolved within them, however powerful the telescope. If the spirals really were composed of stars, they must be very far away. It would be many years before either question could be settled.

In the late nineteenth century, telescopes were combined with photography – the first recorded image of a star was a daguerreotype of Vega, taken on a 15 inch refractor at the Harvard College Observatory only five years after Rosse had completed his 72 inch giant – and telescope design itself became increasingly sophisticated in a variety of ways, except in the one matter of size. As it turned out, that or at any rate the capacity to gather more light, was the key to progress beyond our own Milky Way Galaxy, indeed, to determining that there actually was a

A first step outside our own Galaxy took observers to the Small Magellanic Cloud, which with its larger neighbour is visible by eye in the southern hemisphere. Estimates of distance could be made by comparing globular clusters of stars in our own galaxy (such as the dense mass on the right) with the larger salt-grain sprinklings across this picture: these are similar clusters in and around the cloud, now known to be about 200,000 light years away from us. Henrietta Leavitt studied variable stars in this same cloud, providing crucial data for Hubble's first great discovery.

Cepheid with a period of, say, ten days could be expected to have the same absolute magnitude. So the fainter (at its maximum) it appeared to an observer on Earth, the further it must be from us. To convert that to a measure of distance it could be compared directly with the prototype, a star of which the distance was known by other means. From his studies Shapley could estimate the size of the Milky Way Galaxy itself. In 1918 he described it as a disc about 200 000 light years in diameter, with the Sun about 60 000 light years from the centre, and with a scattered halo of globular clusters above and below the plane. Although somewhat too big, this measurement was a great achievement, but it led Shapley into error. Because of the enormous distances involved, he thought the Milky Way and its disc must be all there was . . . stars, gas clouds, spirals, the lot – for 'Galaxy' read 'Universe'.

Unfortunately for Shapley, the trend of other arguments was

against him. While it was just possible to imagine spirals as incipient solar systems, it was far more appealing to regard them as more distant islands in space. This developed into astronomy's 'Great Debate', one of the major disputes of the scientific world (there was even a somewhat unsatisfactory public confrontation, followed by further salvoes of scientific papers from both sides). No clear conclusion could be reached and the controversy rumbled on as Shapley left Mount Wilson to others and settled into an apparently superior job at Harvard, where he served out the remainder of a conventionally distinguished career. As it happened, Harvard's principal contribution to this story had already emerged, in the form of Henrietta Leavitt's Cepheids – but they were yet to make their full impact.

Hubble and Andromeda

Up on Mount Wilson, a young astronomer called Edwin P. Hubble had kept himself out of the debate. He was already pretty sure of the answer: they would, indeed, be galaxies. Train a 100 inch telescope on the biggest spiral nebulae and it should resolve the individual stars.

Hubble, like Shapley, was born in Missouri and had started out in a different field. Hubble had studied law as a Rhodes Scholar at Oxford – where he developed an English accent strong enough to irritate any who took a dislike to his remote and autocratic style – then switched to astronomy soon after his return. He served his apprenticeship at Yerkes and, having got himself appointed to Mount Wilson by Hale, went off to fight in the First World War. When he actually took up his post in 1919, the new telescope was ready for him.

His approach to the problem of nebulae was systematic. He first studied and classified the amorphous nebulae that might be expected to be part of our own galaxy, distinguishing them from the globular clusters and spirals. He worked on what has since been classified as a small elliptical galaxy that is much closer to us than the giant Andromeda and Triangulum nebulae, resolving its individual stars, including many Cepheids. From the distances that these gave him, this 'object' appeared to Hubble to be surprisingly far away. Then he moved out to the spirals: could he see stars in those, too?

There was one kind of 'event' that he certainly could recognise as being associated with stars. From time to time he saw the sudden, new, sharp pinpoints of light called novae,

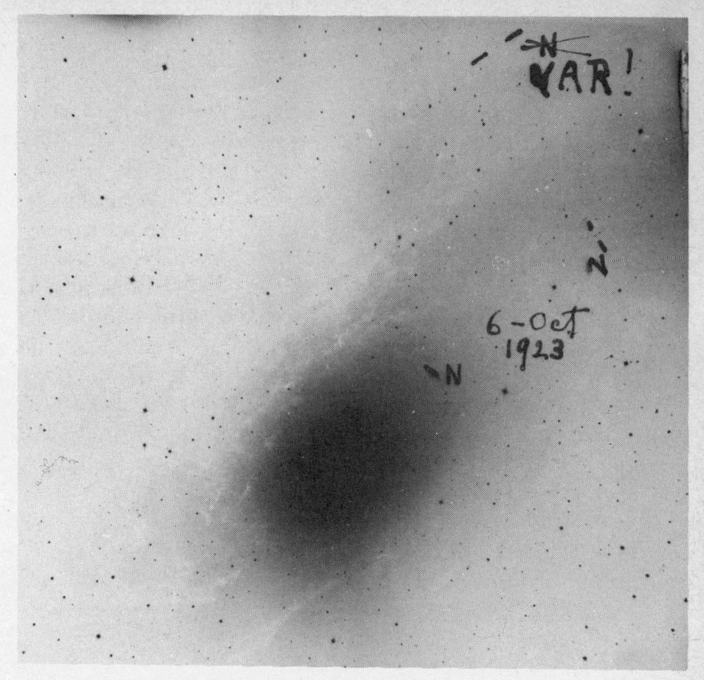

Hubble's record of his crucial discovery of a Cepheid variable in the Andromeda Galaxy.

where unstable stars had abruptly blown off a glowing shell of matter. Their name originally meant 'new stars', which in fact they are not, and as heavenly events, they are much more common than supernovae, in which complete stars disintegrate to make a far more brilliant flare.

In 1923 Hubble had marked several suspected novae with a black-inked 'N' on photographs of Andromeda, when one that had died away began to brighten up again. He realised that this was not a nova but a Cepheid variable. It was just the evidence he needed. He triumphantly crossed out the 'N' and marked it with the new legend 'VAR!' in red ink.

Why red? Allen Sandage, his successor at the Hale Observatories grinned at that question and answered, 'Because he knew you were coming.' Hubble, though aloof and at times unapproachable, was not entirely unresponsive to the needs of a posterity that would gaze upon his work.

Hubble's calculations located the Andromeda Nebula

900 000 light years away from us, well clear of Shapley's limits for the Milky Way Galaxy and impossible to interpret as an outer suburb of it. Even so, Hubble was surprisingly slow to publish and simply went on with his work, gathering more and more data. Other astronomers continued the Great Debate throughout 1924 and only in January 1925 was a letter extracted from Hubble and read before the American Association for the Advancement of Science. It instantly won him a cash prize, and settled the question of 'island universes' for all time, though some delay continued. Scientific papers on the nebula in Triangulum, which produced even more data than Andromeda, and on Andromeda itself did not appear in print until 1926 and 1928. With only one 100 inch telescope in the world, there seemed to be no competition. In fact, Shapley himself could have made the discovery long before if he had asked himself the right question and looked in the appropriate place because, as it later turned out, the Cepheid that Hubble found could also be seen by the less powerful 60 inch that Shapley had used.

Throughout his life, Hubble continued to call the spirals 'nebulae'. The term had scientific respectability. 'Galaxies' was too romantic for him.

Red Shift

While Hubble was setting out to demonstrate the distance of a handful of galaxies, Vesto Slipher, director of the Lowell Observatory in Arizona, was investigating their angular velocities (are they truly rotating systems, as their spiral forms suggest?) and their radial movement (are they moving towards us, or away?). Slipher was an expert spectroscopist, and also used certain peculiarities of the absorption-line spectra of galaxies to indicate motion. In the laboratory, the patterns of lines that were characteristic of particular chemicals always sat at the same points among the colours of the spectrum. In the spectra of the spiral nebulae they would usually be shifted a little to one side or the other.

Pure colours are, in fact, different wavelengths within a narrow region of the electromagnetic spectrum. Beyond visible light in one direction, as the wavelengths get shorter, are ultraviolet, x-rays and gamma rays. In the other direction – longer wavelengths – are infrared and, eventually, radio. Spectral lines, too, exist outside the visible region.

When the source of a waveform moves, the waves get bunched

together in the direction in which it is moving and open out in the direction from which it is moving away. A train may pass through a station at constant speed, but if it whistles as it goes through, to a listener on the platform the pitch of the whistle will start high (shorter sound waves) and end low (longer waves). It is the same with electromagnetic waves. Electromagnetic radiation from an object approaching us is raised in pitch, which means that it is shifted fractionally towards the blue, ultraviolet and x-ray end of the spectrum. Radiation from a departing object is lowered in pitch, shifted towards the red, infrared and radio – an effect that is called red shift, for short.

Slipher found that in spirals that are seen edge-on, or partly so, one side is moving away and the other towards us, typically at speeds of 200 kilometres per second. Also, except near the centre, much the same rotational velocity is observed at all distances along the radius. So the outer stars take much longer to go round the hub than those closer in, which offers an explanation for the winding of the spirals. Another effect seen in galaxies is that the spectral lines appear as broader smudges than are seen in the spectra of individual stars. This is because a galaxy is composed of many stars travelling at different speeds, or in different directions relative to us.

Slipher also saw that there was usually a bodily displacement of the whole system of broadened lines. It would be a very tiny, but measurable shift. In a few, including Andromeda, the spectrum was shifted towards the blue; they were getting closer to us. Most galaxies showed red shift: they were moving away.

Hubble's Universe

Hubble, with the 100 inch, could apply Slipher's method to much fainter objects, but in those days photographic plates were not very sensitive, so exposure had to be long. For any observation, the telescope was first turned to the co-ordinates of the object of the night and the image centred by eye. Just outside of the required photographic field the position of some other object, a reference star, was noted and an eyepiece set to follow that. The photographic plate was locked into place in the focal plane and the shutter opened. Then the observer had to ensure that the reference star remained centred on the crosswires for as long as it took to expose the plate. That might take the rest of the night or, sometimes, the rest of the week.

Hubble wanted to look, stage by stage, as deep into space as

 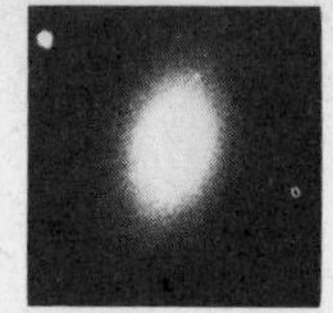  

Ellipses – ranging from spherical to lenticular.

  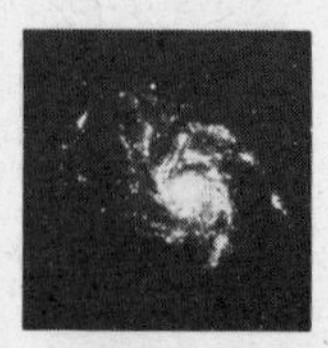

Spirals – becoming increasingly open.

  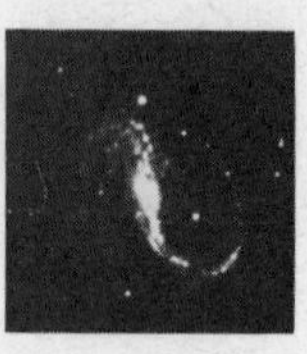

Barred spirals – variants of the more usual spiral shapes.

 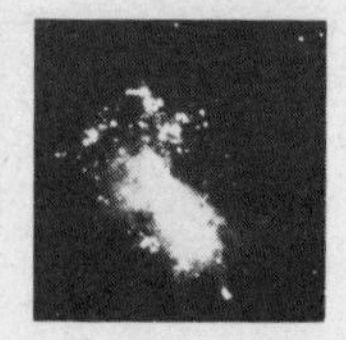

Irregular galaxies – difficult to classify.

Above: *Hubble's Universe – upon which he hoped to impose an evolutionary order based on his system of classification.*

he could; he wanted to make counts of galaxies in different directions, and he wanted to classify all the different kinds of objects that he saw. His system of classification, with elliptical galaxies at one end and open spirals at the other, has endured mainly as a convenient descriptive system, but his other work again struck gold. To collect all his data, Hubble needed help with the donkey work, and that was provided by one Milton L. Humason, who had graduated from mule-train driver to become

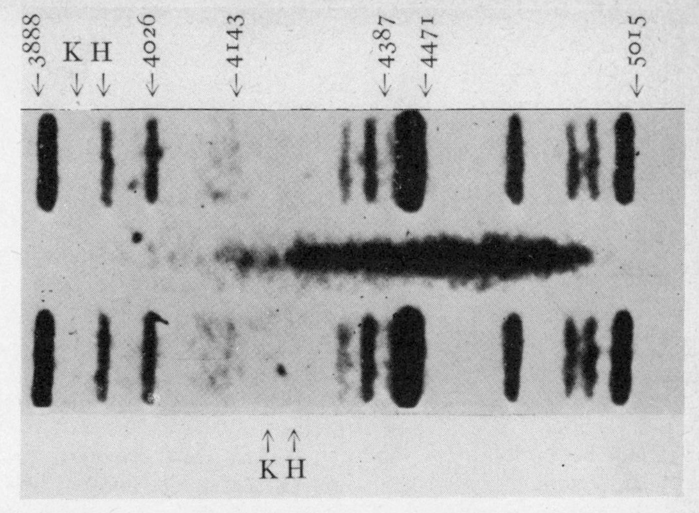

A temporary edge to Hubble's Universe. For many years, with Humason's observational help, he recorded his march across extragalactic space with markers such as this. The three faintly-striped streaks are spectra. The inner smudge is produced by light from the distant object under study, and its displacement (redshift) relative to the reference spectra indicated 'apparent velocities' – or distance across an expanding universe. The original black and white image occupied a few millimetres of a microscopic slide.

a first-class astronomer in his own right, sharing in Hubble's discoveries and writing scientific papers of his own.

Nicholas Mayall (who later became director of great observatories, and in his retirement has a much larger telescope named after him) was a graduate student in 1929 and worked with Humason. Mayall treasures the memory of a particular occasion when Humason developed a plate that had taken a week to expose, a tense and dramatic event in its own right, after such an investment of effort. Soon after dawn the plate was ready for examination. It was a piece of glass the size of a microscope slide, with a tiny black smudge at the centre. Humason placed it on a frosted glass light-box and peered at it with an eyeglass.

'My God, Nick,' he exclaimed, 'we've got a whopper!'

It was the red shifted spectrum of a galaxy that was flying away at apparently unprecedented speed.

Humason picked up the telephone and Mayall listened on an

extension as the news was passed to Hubble, down in Pasadena.

'Mr Hubble,' Humason said (this was his customary form of address), 'you are right again, you hit it right on the nose. We have twice as large a red shift, and your distances must be very good.'

There was silence on the line.

Then Hubble replied, 'Well, Mil, now you're beginning to use the hundred inch the way it should be used.'

And this was enough to send the elated observers off to celebrate with a bottle of what Humason called 'panther-juice', a curious transparent liquid that had a bite and flavour that was new to Mayall. It was gin.

Mount Wilson's Golden Age

The papers that followed, published in Hale's own *Astrophysical Journal*, described a Universe composed of a vast number of galaxies distributed throughout the whole of the space that the 100 inch could reach. The fainter, and therefore the more distant that they appeared to be, the faster was their apparent speed of recession. Hubble had estimates of the actual distance only of relatively nearby galaxies, but from these he worked out a correlation between red shift and distance, since called the Hubble Constant. Although Hubble himself was cautious about the connection between red shift and velocity – he always wrote of 'apparent velocities' – the implied expansion of the Universe provided powerful support for a theory of its origin and evolution described as the 'big bang'. Extrapolating back from the rates at which Hubble's galaxies appeared to be flying apart, it became possible to calculate how long ago the 'big bang' was. Over the years, most estimates have been in the range of ten to twenty thousand million years.

There have been competitors to 'big bang' cosmology – this was another debate that continued over many years – but for followers of Hubble, there remained only details to be confirmed: increasing refinement of the actual figure for the Hubble Constant, and a second number that would tell how the expansion of the Universe would slow down under the influence of gravitation acting between all the matter within it. This, in turn, depended on some measurement of the density of matter. To put that another way, would the Universe go on expanding for ever; was it 'open'? Or would it eventually fall back in on itself; was there enough matter to 'close' the Universe?

A major correction to the distance-scale and, by extension of the argument, also to the age of the Universe, came from an astronomer named Walter Baade. By the late 1930s, it was becoming increasingly difficult to observe faint objects from Mount Wilson because, as Los Angeles grew, more of its nightly glow was reflected back into the telescope from the sky above. But Baade had a splendid stroke of luck. During the Second World War they dimmed the city lights. And that was not all: while most of his colleagues went off to fight, Baade, as a German citizen, stayed behind and was allowed to continue with his astronomy.

He studied the Cepheids in Andromeda and found a curious anomaly: there were, it appears, two different kinds of Cepheids. For although all the stars in the Andromeda Galaxy must be much the same distance from us, the period-luminosity relationship did not have one single value, but two, and these tended to occur in different regions of the galaxy. Indeed, the star populations as a whole had different characters in different regions. 'Population I', in the spiral arms, stars like those near our own Sun, were really not very much like the 'Population II' stars in the galactic centre. This work, published in 1952, showed that the Milky Way and Andromeda were galaxies of comparable size, and therefore doubled the distance for Andromeda. There was a lot more space in Space, and the Universe was much older and bigger than had previously been thought.

On the more local scale, it is now clear that the Population II stars are ancient, long-lived first generation stars, while the others are much younger. The Population I regions contain stars that have a lot more of the chemical elements that are essential to life. It is no accident that we live in a spiral arm, but how that came about is a couple of other stories.

Palomar

By the early 1930s Hale and Hubble had decided they needed to collect more light, with less of it from Los Angeles. With surprising ease, Hubble's reputation attracted pledges for six million dollars for new telescopes on Mount Palomar, hidden deep within the mountain ranges to the southeast of the Los Angeles basin. One was the great 200 inch, a vast instrument for which the mirror alone weighs sixteen tons. It took a year for the glass casting to cool and another to grind the fine 'figure' of its surface.

Hale and Hubble also wanted to chart the galaxies in as much of the sky as they could see from this site. The 200 inch was useless for that: to examine the whole sky systematically would take a thousand years, so narrow is the needlelike shaft that such a telescope views. So they also built a broad field sky-camera of a design invented by an Estonian, Bernhard Schmidt, in 1930. The Schmidt telescope could photograph regions of the sky rather more than six degrees across, recording them on plates measuring about fourteen inches on each side. Such plates now provide the standard reference survey of the northern skies. On it the Moon would be the size of a tenpenny piece, with Andromeda very much larger. Looking with an eyeglass, vast numbers of the fuzzy images of galaxies can be seen almost everywhere on those of the plates that are not dominated by foreground stars or gas clouds. Even with the Schmidt, the first full survey took five years for the hundreds of plates of sufficient quality to be collected.

The team was joined during this time by George Abell, who was particularly interested by the erratic distribution of galaxies in space. The Milky Way and Andromeda had their companions. Both groups appear to be in the outer suburbs of a rich cluster of galaxies that is spread beyond the constellation of Virgo; another prominent one of the Schmidt plates is in Coma and is much denser still. Abell catalogued almost three thousand such clusters, which are themselves unevenly distributed: it turns out that there are clusters of clusters, or superclusters, too.

This is well illustrated by a map showing the distribution of the brightest million galaxies that can be seen from the Northern Hemisphere of Earth. It was prepared by a group at Princeton from a smaller Schmidt telescope at Lick Observatory. A very irregular layout is revealed, with dense congregations, filamentary strands and great holes of emptiness. It appears that a million is about the right number to reveal this degree of structure. Count only the most prominent and the sky seems dominated by a few nearby clusters with pronounced gaps

Right: *The distribution of galaxies in space: a computer-generated map based on a count of the million brightest, as seen from a Schmidt telescope at Lick Observatory. The field is divided into tiny squares which range in tone from black (no galaxies visible) to white (ten or more). The Milky Way is near the outer edge of a cluster centred in Virgo; the huge, dense Coma Cluster (near the 'North Pole' of our own galaxy's rotation) is much farther away.*

GALACTIC
NORTH
POLE
SERPENS VIRGO CLOUD

between them; count a great many more, which is impractical other than for small regions of the sky, and the distribution becomes progressively more even. As Abell comments:

> Superclusters seem to be the largest units in the Universe. Over extremely large scales compared to the sizes of superclusters, we find that one part of the Universe seems to be about the same as any other. Space is remarkably homogeneous. It is what we call the cosmological principle: the uniformity of the Universe on the large scale.

And that is where the optical telescope, looking at the galaxies discovered by Hubble, leaves us. As far as we look back in time, even halfway back to the 'big bang', galaxies do not seem to have evolved much, if at all. It needed, and took, a completely new technique to open up fresh questions and find further answers.

The Radio Revolution

The whole of early astronomy was built on the evidence supplied through a narrow channel of the electromagnetic spectrum, a single octave in a scale that extends like an elongated piano keyboard into the mists on either side. The confinement of early observations to that one octave at which astronomers' eyes work is not quite so arbitrary as it may appear. Evolution has given us eyes that see at these wavelengths because Earth's atmosphere admits as much of the Sun's radiation as lies within that range, but is relatively opaque to wavelengths that are shorter and longer.

There is one exception to that general rule, one other waveband which reaches Earth. It does not carry a great deal of energy from the Sun, and so would not illuminate our surroundings brightly even if we could see it, but it also exists as a second 'window' in the atmosphere through which we can observe the heavens. It is the range that we use for radio transmission.

By using radio, would we just 'see' more of the same story as before? (Or in the case of many stars, like the Sun, the same, but less?) Or would the Universe appear in an entirely different cast? In retrospect we can see that a revolution has taken place, but in the early days radio astronomers had difficulty in persuading 'real' astronomers to take their infant science seriously, when all they could show for their efforts was a squiggle on a piece of paper or a hissing noise from a loudspeaker. There was no image of a kind that optical astronomers were used to, although a spectrum with the usual kind of lines was soon shown to be possible.

Today, detailed radio images can be constructed. To obtain images of galaxies equivalent to those from the optical 200 inch, a Y-shaped array of radio dishes (receivers) has been spread over nearly twenty miles of New Mexico desert. In eight hours the array rotates with Earth to fill in gaps, so creating a time-lapse impression of a huge eye staring into space. Twenty-seven individual elements follow and point at the chosen object in unison, under the control of one computer; a much larger computer extracts a signal from all the noise (for technical reasons, an enormously complex task); and a third converts it into an image, a picture that the human eye can hope to understand.

This has taken over thirty years of development since radio astronomy became a growing field of study. In the 1950s and 1960s there was an extraordinary era of radio telescope building in Britain, Holland, the United States, Australia and many other countries. Beneath Europe's cloudy skies astronomy had languished since the capital cost of big optical telescopes had made the efficient use of their time so important. Now, with instruments that could be used by day and night and in any weather, a new age began. There were two main lines of development: towards big dishes like the fully steerable Jodrell Bank Mark I, designed to receive the strongest possible signal, and the arrays of smaller dishes that are designed to assemble an image. These latter can in any case add up to the same collecting power as one large dish, though the most sensitive radiotelescope in the world is still a single bowl 1000 feet in diameter, formed within a natural valley among hills near Arecibo, in Puerto Rico.

Because it was so difficult to pinpoint their position, the earliest conspicuous radio-emitting objects received new names, pending their optical identification. The first galaxies to be observed were and still are commonly known by their radio names, Cygnus A and Cassiopeia A, the most prominent radio signals in those constellations. Two more identified at once with well-known visible galaxies were Virgo A and Centaurus A. Their characteristics added a new category to Hubble's classificatory system – they were called 'peculiar'.

The radio images of these galaxies had huge 'jets', puffs of matter of some kind that were emerging from them. The galaxy M-87, the optical counterpart of Virgo A, also has one obvious visible jet; another balancing one, angled away from us, is less

This spectacular visible nucleus at the centre of the powerful radio source, Centaurus-A. Two radio-emitting lobes extend as vast jets along its polar axes, suggesting violent events long ago at its nucleus.

easily observed. Where there was evidence of any disc, if only in the form of equatorial dust lanes around a fuzzy near-spherical blob, it appeared that the jets were aligned along some kind of polar axis. Sometimes the object sending out the jets appeared to be moving through space, leaving curled jet trails. Radio astronomers also discovered resolvable radio images of the spirals and hubs of normal galaxies. Early examples were the Andromeda and Whirlpool galaxies. The hub of our own Milky Way, being relatively close to us, emitted an apparently strong signal; the nearby spiral galaxies were much weaker.

It was only the peculiar radio galaxies that could be observed at greater distances, as measured by the red shifts of their spectral lines. Indeed, some radio galaxies were powerful enough to be observed to greater red shifts than optical telescopes could reach – in other words, further into the past – but again, there was no outer boundary. As they looked deeper, there was again, always, more of the same. With radio, astronomers had extended their view of the Universe, but had also revealed an increased range of problems to be solved.

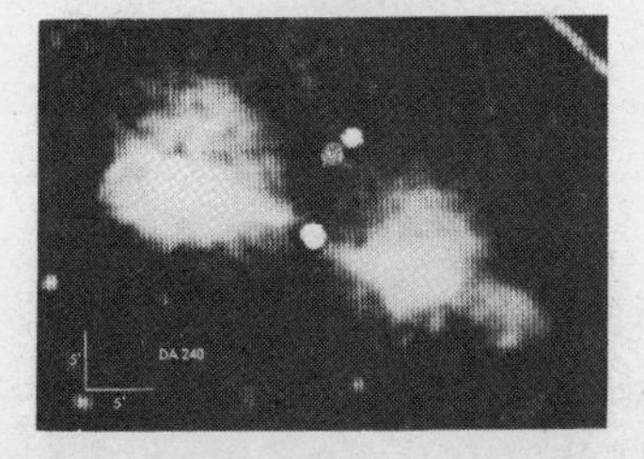

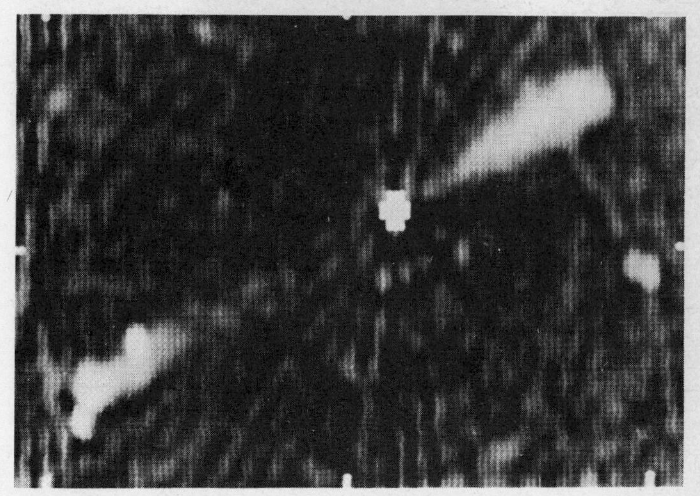

The radio jets of Centaurus-A extend ten times the diameter of the moon across our sky; in comparison, two others though more distant appear far greater still. Unimaginable amounts of energy are required; rotating black holes are proposed.

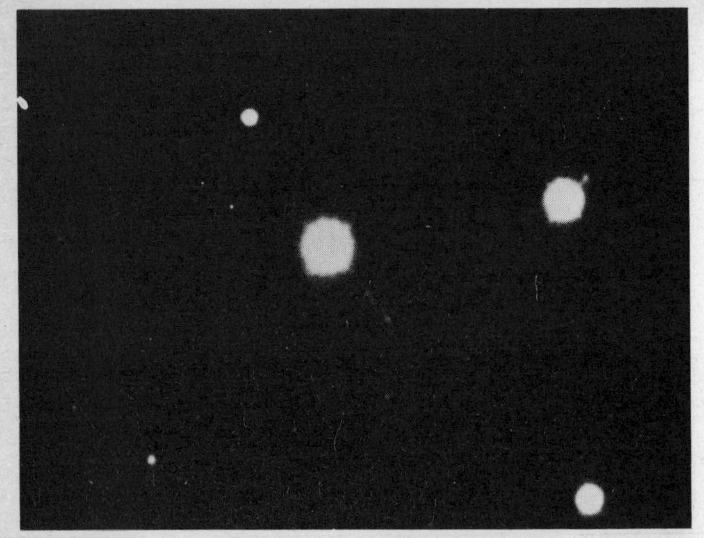

3C273 is a bright radio source for which the optical counterpart, a blue 'object' with a faint visible jet, had an extremely high red shift – which, if cosmological, made it so far away that its energy output was, to some, unbelievably high. A new debate began. Were such quasars nearby, but of nature unknown, or truly 'cosmological' – more frequently found towards the edge of the universe, ie early in its history?

Quasars

Some radio sources proved difficult to identify optically. They were as sharply defined as any radio telescope could resolve, but coincided with no obvious galaxy or star. Eventually, in 1963, in Australia, Cyril Hazard and his colleagues identified the optical counterpart of one of them with the help of the Moon. As the Moon crossed in front of the radio source the signal was suddenly blotted out and then was equally abruptly revealed, so defining a pair of intersecting arcs, formed by the Moon's edges at the crucial times. This gave two possible positions for the actual source and at one of them optical astronomers could see a faint, blue starlike object. At Palomar, Marten Schmidt and Bev Oke observed its spectrum and at first could make no sense of it. The lines did not seem to correspond with any known chemistry. What was going on?

The answer was staggering. The lines could be explained by proposing that they were so strongly red shifted that the source would be travelling away from us at at a sixth of the speed of light

– and, by Hubble's Law, it also meant that the object was 3000 million light years away – which in turn required that it must be extremely powerful, emitting much more energy than would be expected from a galaxy and, as later studies confirmed, from a very small region.

Suddenly a new race was on. For optical astronomers it all depended on whom they knew, as the radio men gained and lost friends depending on the persons to whom they passed their co-ordinates. Now the 200 inch really came into its own and Marten Schmidt became the leading student of these quasi-stellar objects, or quasars as they were called for short.

Theoreticians were confronted with an awkward choice. If the red shifts of the quasars truly indicated their distance, then their fantastic power required that a new source of energy be found, because ordinary nuclear fuel (mainly hydrogen fusion) would soon burn out. Alternatively, a new explanation for red shift was required. In a sense, this new debate was a variant on the older one in which it was questioned whether Hubble's red shifts do truly indicate that the Universe is expanding. The quasar controversy has long been weighted towards the view that the red shifts really are cosmological, that is, that they do represent distance, but it has continued to smoulder on over the years.

One powerful piece of evidence was painstakingly assembled by Marten Schmidt. Searching deeper than anyone else could, he found that quasars were unevenly distributed in their red shifts. Given that red shift does correspond to distance and therefore age, there were more in the distant past until, eventually, a distance was reached beyond which no more could be seen. It looked like a 'wall' of quasars.

Turn this picture round, and it could be explained by supposing that quasars were not formed until some time after the 'big bang'; that then a substantial number appeared; and that in more recent times there have been less, perhaps because some have died or otherwise hidden themselves away. This scheme seemed reasonable, provided that several questions could be answered. What was the quasar's energy source? And how were quasars related to optical and radio galaxies? Tentative answers to these questions again awaited the arrival of new techniques.

The Southern Schmidt joins the Hunt

A quarter of a century elapsed before the Southern Hemisphere had a counterpart to the Palomar Schmidt, to complete the whole sky survey. The UK Schmidt, an outpost of the Royal Observatory at Edinburgh, was built at Coonabarabran in the Warrumbungle Mountains of New South Wales, alongside another new, big instrument, the Anglo-Australian Telescope.

In design, the UK Schmidt was closely based on its northern brother, with minor but significant improvements. One is in its optics: Schmidts have not only a reflector (in these two it is the size of Rosse's 72 inch) but also, at the top of the tube, a smaller objective lens (48 inches). In the Palomar Schmidt this is a single piece of glass; but for the newer telescope it was possible to grind two pieces of glass of this size to fit together and so to provide a substantial measure of colour correction. A second improvement has been in photographic emulsions, which can now be hypersensitised. This involves baking them first in hydrogen gas, then in nitrogen, just before use, driving out contaminating oxygen and water, both of which reduce sensitivity a great deal. The consequence of these improvements is that the southern plates go substantially 'deeper' than those in the north, taking in a greater volume of space.

To 'see' quasars an objective prism is fitted to the Schmidt. This is a further piece of glass at the top of the tube: it has two plane surfaces that are slightly inclined to each other, forming a prism which, though much flatter than usual, will still spread the image of star, galaxy or quasar to produce its spectrum. Each object on the photographic plate will appear not as a dot or blob, but will be elongated into a rainbow dash of colours. In practice, however, black and white plates are used: the astronomers can work out what the colours would have been.

Quasars have characteristic spectra, with very bright emission lines from glowing clouds of ionised gas around their powerful energy source, and these are relatively easy to spot in the Schmidt plates. Typically each plate (covering a region of sky about 6° square) produces one or two, a small number compared with all the galaxies. By all techniques about 1500 had been found by 1980 and they appeared to be spread fairly evenly throughout the sky.

The Anglo-Australian Telescope, the southern Schmidt's big neighbour, is an early example of a far bigger revolution in astronomy. Although at 162 inches it was rather smaller than the Palomar 200 inch and otherwise rather conventional in layout, it was designed from the outset to have a much greater collecting power. The trick was to use electronics rather than photography. Even with recent improvements, photographic emulsion catches only about 1 per cent of the light, comparing very unfavourably with the human eye, as employed by Rosse. Electronic detectors are much more efficient: they aim to register every incoming photon. In practice it is not too difficult to pick up two out of three.

In the Anglo-Australian Telescope the light is gathered by the big reflector and sent up to the top of the tube, where it is generally returned downward by a secondary mirror through a hole in the primary to a cage that hangs below it: here is what is called the Cassegrain focus of the telescope. The AAT's Cassegrain cage is crammed with electronics and the observer is banished to a room nearby where he watches the Universe by television.

The traditional way of observing was to spend the night on the telescope itself, muffled against the mountain cold inside the open dome, checking from time to time by eye that the image had not drifted off the cross-hairs of the sighting telescope. Today the astronomer observes in air-conditioned comfort at a computer terminal. Some older hands feel that the romance has gone out of observing the Universe; that the physicists are taking over. Others revel in the power and responsiveness of the vast machine at their fingertips, feeling an expanded awareness of the Universe, as the telescope's electronically enhanced senses delicately touch object after object in the skies. They can choose to see farther and fainter or, for the brighter sources, examine many more in a night.

The telescope has the full-time use of one of three sets of equipment designed by Alec Boksenberg of University College, London. Light from the telescope is fed through a four-stage image-intensifier that increases the strength of the image ten millionfold so that in the video picture that emerges each original photon can be represented by a dot. In the faint image of a galaxy, each field of a fraction of a second may regiser only a few of these dots but, if they are stored in a computer, a detailed picture will gradually build up. The computer can also process the information to other kinds of display – for example, the spectral spread of 'colours' present in a designated slice of the original picture. In addition, the intensity range can be changed at the touch of a switch to consider in turn the brighter or fainter components of a display.

With another set of his equipment packed into crates, Bokesenberg travels the world to set it up on other big telescopes that were designed before electronics began to take over.

M87 is a radio galaxy (Virgo A) with its nearer jet plainly visible in the shorter-exposure inset. When analysed by electronic techniques, the central region of M87 is found to be darker than expected – a black hole, perhaps.

'Boksenberg's Flying Circus' is welcomed to most of the bigger American telescopes. As the resulting scientific papers have local as well as visitors' names on them, this is an arrangement that benefits both sides.

Black Holes

The galaxy M-87 is a strong source of radio and x-rays as well as being a prominent, large, elliptical galaxy, as seen by its visible light. As mentioned earlier, it also has jets, one of which can also be seen optically. Plainly, this is a different kind of galaxy from our own – but how different? When Alec Boksenberg investigated it, what emerged from his study well justified the designation 'peculiar'. Certain large, yellow stars have a visible spectrum that has very obvious dark lines that are due to the absorption of light by calcium atoms. M-87 has so many of these stars that the calcium lines stand out clearly in the spectrum of the galaxy, too, and random movement of individual stars relative to the galaxy as a whole broadens the lines into smudges. More detailed analysis shows that the broadening of the lines increases markedly towards the centre of the galaxy: it is like a cloud of flies in which those near the middle are buzzing about much more frantically than those farther out. Further, although in general there is an increase in brightness towards the middle, the centre is not in fact quite as bright as might be expected. This provides strong evidence for a black hole.

Since they were shown to be theoretically possible, black holes have been invoked as possible explanations of a wide range of odd or unexplained astronomical phenomena. According to present-day physics, a sufficient concentration of matter could cause the gravity drawing its substance inwards to be greater than the forces that hold it apart. Once inside this boundary further collapse could not be stopped by any known physical agency and even light itself could not escape – hence the name. Physicists have calculated that the energy released by matter falling into a black hole could be greater than that produced by nuclear processes by a factor of hundreds or thousands – enough, in fact, to account for the brilliance of quasars if they really do lie at cosmological distances.

M-87, though enormously energetic as galaxies go, is not quite in that category, but to Boksenberg it strongly suggested an evolutionary sequence. Peter Young, a young British astronomer at the California Institute of Technology, who worked with Boksenberg on M-87, suspects that there would be a black hole at the middle of nearly every large galaxy. He suggests that at one time, every galaxy would have been a quasar, but now as the density of the Universe decreases with time, the rate at which black holes are to be found has gone down. So the quasars have died out; they have been starved of mass, and we no longer see active quasars but only dead remnants. At this late stage the black hole would lie quiescent at the heart of a galaxy whose remaining substance mostly had motion enough to evade capture. Could there be a black hole even within the star clouds at the heart of our own galaxy?

Boksenberg has also observed another quasar, Ton 256, which by his imaging technique he has shown to have the fuzzy, extended starfield of a galaxy about it. But more commonly, a distant quasar could be expected to outshine its galactic halo a thousandfold; such quasars are in any case at distances where galaxies themselves are too faint to be seen. The radiogalaxies with prominent jets also require enormously powerful sources of energy; and where there are repeated 'puffs' of matter along the same alignment, they demand a source that can 'remember' its orientation for aeons. A massive, rotating black hole fits that description, too, as little else does.

The Double Quasar: a Gravitational Lens?

It was radio astronomy that led to the discovery of a quasar which, on the face of it, had very strange properties indeed. For about seven years, a team led by Denis Walsh at Jodrell Bank had been using their big dish to make an accurate survey of the position of radio sources, in order to find the corresponding optical objects. They had already found hundreds, including many galaxies and quasars, when they came across one that appeared to have two members separated by one six-hundredth of a degree, much closer than would be expected by a chance line-of-sight coincidence.

At the Kitt Peak National Observatory in Arizona, where Boksenberg's M-87 study was also made, Walsh and the others began studies of the spectra of the two quasar images which showed that they were virtually identical, with the same emission and absorption lines and the same red shifts. It soon became clear that there was no reasonable alternative to the conclusion that the two images were of a single quasar, seen by separate paths through the sky. A large elliptical galaxy, about halfway between the quasar and the Milky Way, was acting as a gravitational lens, bending the quasar's rays in such a way that it could be seen twice.

On a Palomar Schmidt plate, blurring by Earth's atmosphere caused the two images to overlap, but on a particularly good night for 'seeing', a telescope at the top of the highest mountain in Hawaii was able to resolve them. Alan Stockton of the University of Hawaii then processed his photograph to eliminate one of the quasar images and reveal the galactic 'lens' that lay between the quasar and Earth. Although it turned out to be the largest in a group of sixty, this galaxy was still much fainter than the quasar that was twice as far away.

At Palomar, too, a few days before Stockton's observation, Peter Young and his colleagues had used the 200 inch with a new kind of electronic detector to resolve the images and went on to complete a description of the whole arrangement and to work out its implications. One is that quasars really must be 'cosmological', that is, a long way away. Another is that Einstein's theory of gravitation applies at cosmological distances as well as in the solar system: this demonstrates that an important law of nature had not changed significantly in time. (Einstein, incidentally, had prediced the possibility of gravitational lenses, but in the Universe as then known did not expect one ever to be found. Now not just one, but two cases have appeared.) As a further bonus, the quasar 'weighs' the intervening galaxy; its mass can be calculated from the way it bends light; and, finally, the geometry of the light-paths gives Young a way of measuring the radius of the Universe.

Young's own future programme is to use the new detector (essentially, a solid-state chip borrowed from the latest television technology) and the 200 inch telescope in a five-year programme to chart a small sector at the faintest limits of the Universe, in order to try to discover how galaxies evolved in their earliest existence as visible objects. Apart from the hints from the quasars, little is known about their origins or even which came first, individual objects such as quasars and galaxies, or clusters or even superclusters of galaxies. It is, however, possible to make mathematical models of what might have happened after galaxies had formed and to trace them forward in time to see how they compare with what we see today.

Gravitation in an Expanding Universe

Between centres of mass that are far apart in space, gravity is the only known force that can have any significant effect. So in a Universe that is expanding, in which galaxies are flying apart, gravity governs their distribution in space. To calculate how a large number of bodies would interact is an inhumanly difficult problem, but lies within the capacity of a big computer, given sufficient time.

At Cambridge University, S. J. Aaseth set up a mathematical model in which a few thousand points distributed randomly throughout a sphere could be set in motion, with the whole system expanding, and allowed to interact. As a model, it is a very limited representation of reality: there are far too few 'objects' in the model and there are no sharply cut off boundaries in the real Universe. But it does give a glimmer of understanding of how galaxies interact. After periods of time equivalent to ten or twenty thousand million years, the distribution does become uneven, with at least a superficial resemblance to the clusters and superclusters of galaxies actually observed.

In further runs the starting conditions were changed a little, and it was found that with just a small decrease in the randomness, that is, a little more order, or structure, at the start, similar clustering could be achieved much faster. At the end, the experiments produced distributions consistent with what we now see, but did not reveal much about how the Universe might have looked when galaxies had first formed, nor even how long ago that was.

Direct investigations into the age of the Universe have continued along the lines established by Hubble fifty years ago. In 1979 and 1980 three young investigators working mainly at Kitt Peak in Arizona were busy changing the numbers yet again. It had been observed that, in rotating galaxies, the faster they go round, the brighter, on average, they seem to be. Marc Aaronson, Jeremy Mould and John Huchra took up this idea and used it as a measure of distance, rather as the Cepheids, which also had two interrelated variables, had been used for the nearest galaxies.

The team selected bright galaxies that were more or less edge-on to Earth, collected their spectra from the most sensitive radio telescope available (the 1000 foot dish at Arecibo) and noted how much a particular prominent line in the spectrum had been broadened out by the rotation, due to one side of the galaxy moving towards us and the other away. At such an orientation, dust clouds in their plane might block off the light from many of their stars, as it does when we look across the plane of our own Milky Way, so they decided to measure the brightness of the galaxies, not in visible light, but in the infrared, which is less affected by dust. Much of the infrared would be lost in Earth's own atmosphere, but this did not matter too much as the main limits to their measurements were, in any case, set by the strength of the radio signals: these were not the 'peculiar', strong radio-emitters, but weaker, ordinary galaxies.

Previous measurements of the rate at which the Universe was expanding had gone as far as the Virgo cluster, the nearest large congregation of galaxies, of which our own Local Group is an outer suburb. Extrapolation to the Universe at large had given figures for its age that varied from about 13 000 million years up to, by some accounts, as much as 20 000 million years. Aaronson, Mould and Huchra agreed fairly closely with earlier results for the Hubble expansion as far as the Virgo cluster, but were able to go farther. Beyond Virgo, they found that there was a jump in the expansion rate, to about half as much again. By the new calcu-

The nearby Virgo cluster of several thousand galaxies (of which this is part of the central region, 70 million light years away) exerts an inward gravitational pull which compensates for the overall expansion of the universe. This may have led to overestimates of the age of the universe: is it younger than hitherto believed?

lation, the age of the Universe would be only about 10 000 million years. The age of some star clusters, calculated by completely different means, is supposed to be far greater than that, so a direct conflict has arisen. However, there have been so many changes already, as astronomical data and ideas have evolved, that more cautious voices comment that to be able to estimate the age of the Universe to a factor of two is itself an outstanding achievement.

Looking at the new results another way, it appears that compared with the more general rate of expansion, Virgo and the Local Group have a component of 'infall' which suggest that they are gravitationally bound. Even if the rest of the Universe continues to expand for ever, many of its component clusters will fall in upon themselves. The forecast, Universewide, is for intergalactic collisions on an increasing scale. Indeed, the skies may already be full of their products.

Colliding Galaxies

At MIT some years ago, applied mathematician Alar Toomre fed data representing pairs of simplified rotating galaxies into a computer, set them in motion towards each other and watched

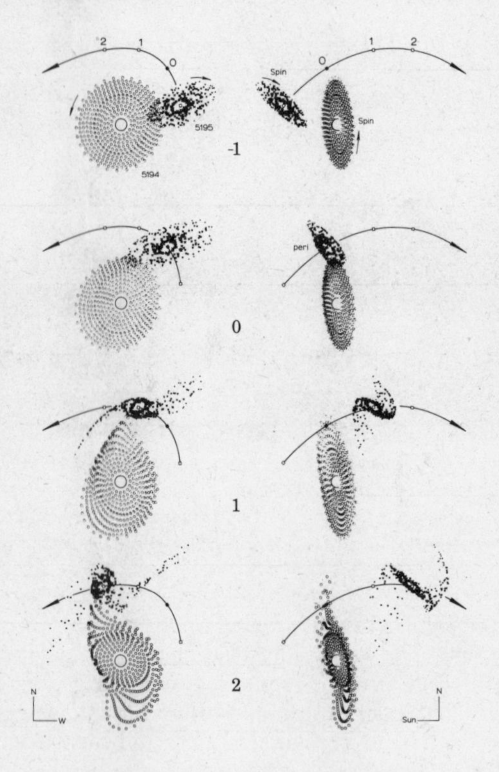

them collide. One interaction had a smaller galaxy pass close by a larger one and in the plane of its spirals. The smaller mass swung round the larger, pulling a spiral arm out to link them, and at one stage they looked remarkably like the Whirlpool pair do now. Two others approached at an angle and also drew spiral streamers out from each other. Again, just such an interaction can be seen in a double galaxy called 'The Antennae'.

Andromeda and the Milky Way are also approaching each other at an angle. But there is no imminent danger: at the time of closest approach, the Universe will be twice its present age. In any case, such events are not a conventional kind of collision. Stars are so far apart in space that chances of impact or even near misses are remote. Radio noise and other radiation from interacting gas clouds would be more noticeable, but not

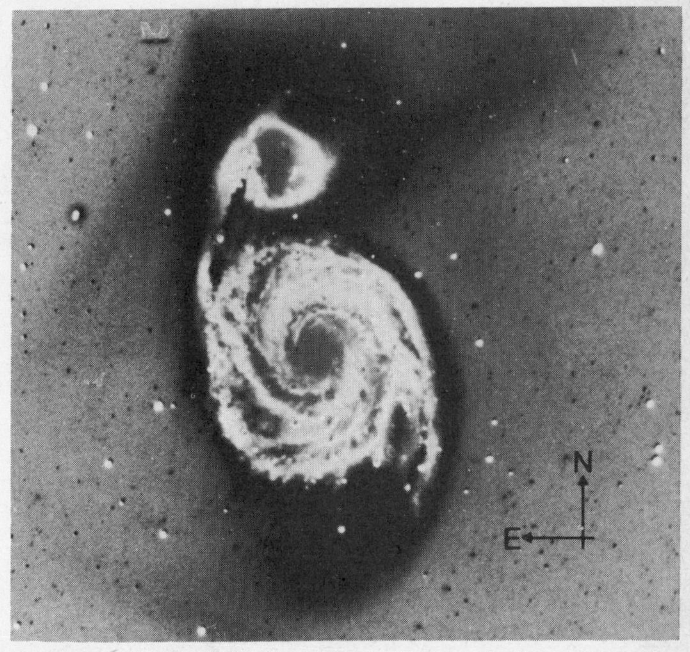

Opposite: *Galaxies colliding in a computer: four scenes (and two angles ending in a distribution of stars (*lower left*) that looks remarkably like M51, Rosse's 'Whirlpool'. The model predicts fainter plumes which were indeed found when a Palomar Schmidt picture* above: *was 'massaged' to reveal its lower density of stars. Seen from a viewpoint to the side of the line from our own sun, the companion would have a barred form.*

dangerous. But then, at most places even in an x-ray galaxy such as M-87, humans could have no great problem of survival: such 'events' are extremely 'dilute', to use the astronomers' terms.

That galaxies do collide is not disputed, but observers and theoreticians have several different theories about what may happen in detail, and even of the role of collisions in the evolution of the Universe. Some point to components of galaxies that have anomalous amounts of spin and talk of 'galactic cannibalism'. Others have suggested that some big elliptical galaxies might be the product of collisions in which much spin has been cancelled out, producing the soft-edged amorphous droplet that we see. This last idea is not widely accepted, but could help to explain why some large elliptical galaxies have so much metal in their spectra – a product that requires considerable stellar evolution within the galaxy.

A Gallery of Galaxies

So, it seems, for all that has already been discovered, we do not know how galaxies began or how they will end. We can see examples today and, with increasing uncertainty, at a range of points into the distant past; but we have no more than the haziest idea of how to construct an evolutionary system of them. The different types of galaxies are almost like different species of animals. Even within species, each individual has its own distinctive qualities, as unlike any other as two human personalities or faces. Is there such a thing as a 'normal' galaxy?

Galaxies with a normal spiral or elliptical structure are seen to have greatly extended realms of stars. Around symmetrical central regions asymmetrical knots or visible jets may appear. In a number of cases, far outside an amorphous centre there is a sharp-edged shell of stars: is that perhaps a shock wave passing through a thin cloud of matter, causing it to condense into stars? Or is it, as Alar Toomre suggests, due to some other 'object' falling in or through the galaxy making denser 'waves' in the stars around it?

Plainly, astronomers have a great deal still to learn of the structure of the Universe, as each new discovery opens up as many new questions as it settles old ones. Again the demand is for more light, clearer detail and spectra, and images from a wider range across the electromagnetic spectrum.

New Instruments for Old

The biggest leap forward in instrument design could come from the space telescope, a 94 inch reflector lifted into Earth orbit by the space shuttle. Below our atmosphere stars twinkle and galaxies smudge, losing crucial detail. The space telescope offers the greatest improvement in resolution since Galileo himself turned a telescope to the skies. It makes images ten times sharper, to reach galaxies that are fifty times fainter than could be seen before.

But even a successful space telescope could not make ground-based telescopes out of date; its task is to complement them. It is short both on time available for all that needs to be done and on collecting power: it is light-starved. The most promising way of answering the astronomer's traditional plea for 'more light!' is to build mountain-top telescopes, not with bigger mirrors, but with more of them. The first of a new generation, set at 8500 feet on Mount Hopkins in Arizona, has six elements each the size of Lord Rosse's 'Leviathan' which add up to the equivalent of 176 inches, making it the third largest in the world at the time that it was built. Even before it was fully commissioned it played an important part in the early story of the double quasar, eliminating an alternative explanation.

Radio has no problem with twinkle: Earth's atmosphere offers no barrier to a good radio image, but only recently has it become possible to achieve one, other than by linking distant instruments that are normally operated separately. Now high resolution radio astronomy is becoming possible on a routine basis both at the VLA and in a development of the Jodrell Bank Observatory the total collecting power of which is still less than that of the VLA, but now with a capacity to pinpoint smaller objects. These extended telescopes will be able to resolve objects down to the size of galaxies or in some cases just their cores.

To study the evolution of galaxies, astronomers will need to compare close optical images with those of distant earlier objects, the light from which has been red shifted right out of the visible range. So in Hawaii the most powerful infrared telescope in the world has been built as another outpost of Edinburgh, at a height of 13700 feet, where losses due to water vapour in Earth's atmosphere are minimised. And high in the Azores will go another British dish, designed to collect millimetre waves, thereby filling a gap between infrared and radio. But from those the big results are yet to come.

A precise understanding of the Universe at large is man's most ambitious quest. Like other evolutionary studies it is historical in nature. The history so far uncovered has, at each stage, been more surprising and a great deal more complex than might have been expected. Human imagination is indeed a powerful faculty, but it requires the discipline of observational evidence. If we had no more than our eyes and unbridled speculation, we might still be hotly debating whether Earth went round the Sun. But our extended human senses have taken us far beyond the Milky Way, to reveal the shadowed outlines of a hidden order and, always, to pose more questions still.

The Contributors

SIMON CAMPBELL-JONES has been a producer since 1967, making mostly science programmes for BBC-TV and WGBH Boston. He was Editor of Horizon from 1976–81.

FISHER DILKE took a Ph.D. in theoretical astronomy at Cambridge University. Since 1974 he has worked for BBC television making a variety of science programmes.

TONY EDWARDS has produced science and natural history films since 1973. An Arts graduate, he has a particular interest in scientific revolutions and paradigm shifts.

DICK GILLING is a graduate in English Literature who has written and produced more than fifty science programmes. They include many for *Horizon* as well as episodes of *The Ascent of Man, The Age of Uncertainty* and *Human Brain*.

STUART HARRIS joined the BBC in 1965 after a degree in electrical engineering at Queen Mary College. Working mainly in the Science Features department, he specialised in films on space flight, and was Science Producer on the BBC's coverage of the Apollo space flights. He is now an independent film-maker in California, where he follows the same interests.

PETER JONES'S career as a documentary film-maker began at Granada Television in 1963. Since moving to the BBC in 1969 he has produced the twelve-part *Landscapes of England* with W. G. Hoskins as well as many scientific documentaries, both dramatised and non-dramatised. In 1979 he became Series Editor of *The World About Us* in the BBC's Natural History Unit.

ALEC NISBETT, read mathematical physics, then joined the BBC in 1953. He has been a senior producer with its Science and Features Department since 1972, and has written and directed many Horizons and Tuesday's Documentaries and produced two-hour specials such as 'The Weather Machine' and 'Key to the Universe'. He has won awards for both films and writing.

Picture Credits:
Bell Telephone Laboratory p. 162; Martin Burkhead, University of Indiana p. 211; Alan Burton p. 105; California Institute of Technology, Palomar Observatory p. 179; Carnegie Institution of Washington, Mount Wilson and Las Campanas Observatories p. 187; CoEvolution Quarterly p. 195; Cornell University p. 125; Fotoarchief Sterrewacht Leiden p. 199; Owen Gingerich, Center for Astrophysics, Cambridge, Mass. p. 115; Harvard Magazine p. 178; Peter Jones p. 160; Kitt Peak National Observatory pp. 181, 200, 204, 209; Lick Observatory pp. 113, 131; D. F. Malin, Anglo-Australian Observatory © 1980 Anglo-Australian Telescope Board p. 198; Dr S. Moorbath, Dept of Geology and Mineralogy, Oxford p. 97; NASA pp. 12, 14, 16, 18, 19, 21, 25, 26; NASA/JPL pp. 39, 40, 54, 63, 64, 65, 67, 69, 71, 76, 77, 81, 84, 86, 87; NASA/Viking News Centre pp. 37, 38, 42–3, 44, 46, 48, 51; Princeton University p. 166; Royal Asiatic Society p. 175; Royal Observatory, Edinburgh p. 185; Science Museum, London p. 181 top, 190; Science Photo Library p. 142 (Dr Steve Gull), 143 (Smithsonian Institution), p. 146 (X-ray Astronomy Group, Leicester University), 153, 155 and 156 (all NASA); Space Frontiers/NASA/JPL p. 57; Alar Toomre, Massachusetts Institute of Technology p. 210.